Artificial Intelligence for Business Applications

Use Artificial Intelligence for Scaling Up Your Business Using AI Marketing Tools

Author: Ethem Mining

Disclaimer Notice.

Please note the information contained within this document is for educational and entertainment purposes only. This book not intended to be a substitute for medical advice. Please consult your health care provider for medical advice and treatment.

Table of Contents

Introduction

The book, *Artificial Intelligence for Business Applications,* focuses on ways in which this technology can help in business growth. The tools discussed in the book deal with marketing. Going through the book, one gets to understand what artificial intelligence is and how it started. One will get to see the changes the technology has gone through over time.

Going through the book, one will learn of the opportunities that artificial intelligence technology brings in the world of business and economies. Apart from growth, the book covers the areas of innovation and productivity. In the book, the reader gets information on the challenges of artificial intelligence.

How one can apply the varied applications in this type of technology is discussed. The book covers the topic of how one can get new customers using the tools. Combining the human element with this technology is addressed. It discusses the effect of artificial intelligence on market research, human resource, and the sales process.

One will find specifics on various artificial intelligence technologies. These include discussions on chatbots and autoresponders. The book gives a roadmap to follow when making a choice on which technology to use. A reader will find practical tips on how to promote their technology of choice. How one can connect the technologies with email is discussed.

The book shares practical tips on how to introduce artificial intelligence into a business setup. The author gives pointers on using the tool of strategy in achieving the incorporation of the technologies in an organization.

Chapter 1: What is Artificial Intelligence?

Artificial intelligence is a combination of computer development and human intelligence. Some refer to it as machine intelligence. Here, the machine carries out tasks attributed to humans. The computer systems within the device are responsible for the actions. The ability of the gadgets to carry out the acts without human intelligence makes them intelligent. AI is the abbreviation of artificial intelligence. It focuses on becoming a copy of the human brain. The use of robots to replace the human interface across industries exemplifies AI in the world today.

Artificial intelligence applications are many transversing industries. AI utilizes the power of algorithms.

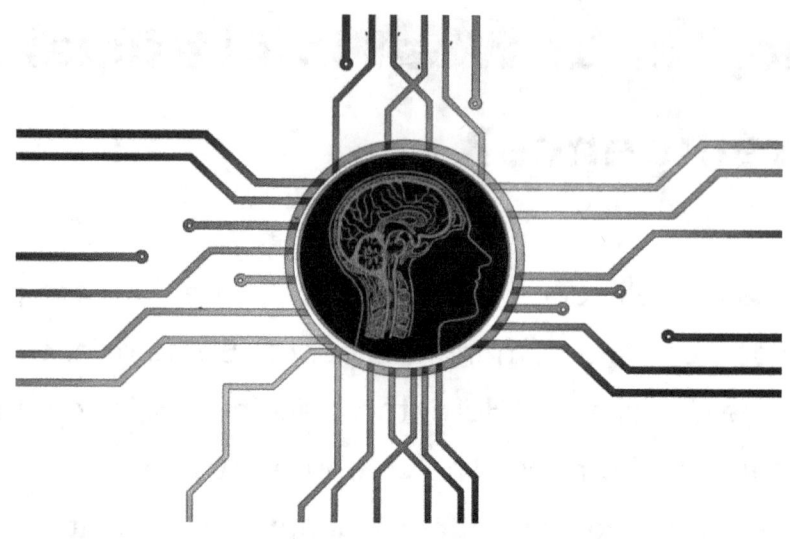

The specific tasks attributed to artificial intelligence include decision making, speech recognition, translation, and visual perception. Planning and forecasting are areas that can utilize AI. There are machine intelligent gadgets that can perform tasks based on previous experience. The actions that the devices carry out range from simple to complex ones. Some robots can multi-task. AI products can now navigate, transcribe, and speak. There are artificially intelligent devices that can recognize emotions either in speech or in what they visually perceive. In the medical field, there are smart devices for diagnosis. The tasks are applicable in many areas, including agriculture, science, and travel.

The tasks attributed to artificial intelligence that involves decision making base themselves on learning, reasoning, and self-correction. Businesses can take advantage of the decision-making ability to determine their way forward. AI allows organizations to process large amounts of data in short periods, therefore, making the decision-making process faster. The decisions made by the devices are considered to be intelligent. The machines base their choices on data. The gadgets learn through inputs. The inputs create experiences for the devices which, allows them to solve problems. These experiences can be considered to be the neural networks of the machines.

With regard to speech recognition, the machines having artificial intelligence convert spoken phrases and words into formats that they can read. The conversion process allows for identification. For the conversion to be effective, the devices should be able to overcome challenges such as accents associated with speech. The use of slang presents another hurdle. The gadget should be able to recognize voices. Speech recognition belongs to the field of computational linguistics. Here, the ability of a machine to predict and utilize input data comes into play.

There are transcription services that rely on this type of artificial intelligence. Voice recognition plus AI gives speech recognition. Depending on the context within which translation occurs, it can be said to be dependent on the speech and voice recognition capabilities of machines. Artificial intelligence can be used to augment the speed and quality of translations carried out by humans. The best machine translators have learned from humans. AI machines are facing challenges when it comes to translating the cultural and emotional aspects of speech. They also face hurdles when it comes to tone, which can change the meaning of a word or phrase. For translation to be effective, AI must link with machine neural network technology.

The visual perception of machines is dependent on how closely they can replicate human visual perception. The better the replication, the more accurate the machine visual perception is. The visual perception achieved affects the intelligence of the machines to a great extent. Machine visual perception helps them make decisions. The ability to see involves the machine understanding what an image is. Visual perception involves deep learning. The stages of machine visual perception are the early stage, intermediate stage, and the high-level stage.

The levels are hierarchal. For the device's visual understanding ability to be effective, it has to be active as opposed to passive in nature.

There are similarities and differences between human and machine intelligence. Most machines are programmed to focus on specific duties, whereas humans can perform a myriad of tasks concurrently. Some devices, like humans, have been shown to have a bias. Lately, a bias factor is added during programming to take care of the bias in machines. Both can be considered to be having neural networks with humans having millions of neurons. These networks help in decision-making. There are tasks that devices can perform faster with a higher level of precision than humans. Human intelligence is considered to be faster as compared to that of machines.

To understand AI, we will look at how it evolved. Here, we will look at how it started, the developments over time, and where it is today. Doing this will give a better appreciation of what it is. We will then expound on the value of machine intelligence, particularly within the business and economic situations. The book will cover the typology of AI applications.

Here, we will delve into how each of them works. We will expound on their differences and similarities. The types discussed are dependent on the type of classification used. Within the chapter, there are discussions on examples of artificial intelligence.

Artificial Intelligence Evolution

To understand the evolution of artificial intelligence, one must understand that AI bases itself on the premise that the human thought process can be mechanized. This kind of thinking goes back to the first millennium. The Chinese and Greeks are thought to have formalized this kind of thinking. Myths and rumors abound during this time of beings from out of the earth that exhibit intelligence. Some societies refer to these beings as aliens. The actual origin of artificial intelligence is unknown.

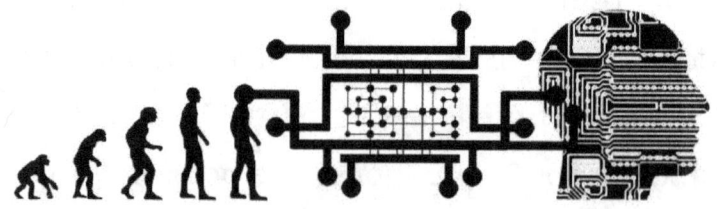

Machines that think first came into light in the 1950s. A Briton by the name Turing created a device that copied human thinking. He published a paper on the same. In the 1800s, there was postulation that machines could think like humans. Alan Turing was part of scientists who believed in devices that could exhibit a copy of human intelligence. He based his creation on the field of mathematics. Claude Shannon was also a scientist who contributed to the subject in the 1950s. The 1950s can, therefore, be seen to be the era where artificial intelligence moved from imagination to reality.

Turing, unfortunately, did not get to see his thoughts accepted into mainstream science. Incorporation of his thought process on machines happened in the mid-1950s. By the year 1959, an AI laboratory was established, referred to as MIT. MIT is an abbreviation for the Massachusetts Institute of Technology. By the 1980s, artificially intelligent machines in the form of personal computers were now available to the common man. Artificial intelligence in the 1990s brings gaming, self-driving cars, and domesticated robots. The 2000s showcase the commercialization of AI devices in industries like marketing and toy manufacture.

AI devices are pitted against humans in the 2010s with robots surpassing humans in performance.

AI origins link to current developments through the Turing test. The test determines whether a machine can think or not. The commonality between the events is data. Currently, studies that are trying to mimic the human brain in both function and structure are ongoing. This kind of learning is known as deep learning. Applications include the prediction of texts and words in messaging or emailing. The focus on further developing facial recognition machines is increasing. Businesses are embracing the use of this technology in workplaces. Organizations in the security industry are already utilizing this technology.

Current developments have brought to the fore concerns on ethics in terms of privacy. Those developing applications that use AI now have to contend with the laws that are coming up regarding this. Chip manufacturers are working on making AI-based features better. Cloud providers are increasing in size as more businesses take advantage of the services they offer. Organizations are using machine learning to make their marketing more effective.

Some AI devices are in use both in military and civilian life. Drones are examples of such machines. Common gadgets in everyday life, like toothbrushes, are now using AI. Current strides made in the field of artificial intelligence are dependent on the ability of devices to learn by themselves.

This type of learning allows gadgets to forecast behavior. Search engines utilize the feature to predict what one is looking for as they type in words. Entertainment organizations currently suggest programs for one to watch using the predictive aspect of artificial intelligence. Applications are now available that can correct mistakes one makes as one is carrying out a task. Medically the machines are quicker than humans in reading biopsies. The advantage of this type of AI is that programming is not required. The role of the developers instead is to ensure that the learning bases itself on correct perceptions.

Some claim that by the year 2050, AI will be smarter than the intelligence of humans on all levels. There are claims that machines will be able to carry out tasks better than humans. The devices will be able to be autonomous.

Their day to day decisions will be independent of humanity. In the future, artificially intelligent machines may replace the human face across industries. AI devices in the future should be able to analyze more information from pictures. Newer software reliant on artificial intelligence will be created to achieve this feat.

The field of automated research will, in the coming days, improve on its use of artificial intelligence. The practice of individualized healthcare with the foundation of the genome is a feature to be seen in the future. Banking institutions will invest more in chips that are AI-enabled. These chips will be optimized. In the future, various AI systems will be learning from each other. As the machines will do this, so will artificial intelligence be able to read and predict global trends. Some companies are working on creating AI machines that will be able to make appointments.

The Value of Artificial Intelligence

One cannot overstate the value of artificial intelligence in the marketplace. Currently, the use of artificial intelligence in core business functions is still low. Organizations that use AI in their day to day operations have found it to be transformational. Artificial intelligence does have its limits. The value derived from AI is dependent on how a company can use it. The impact of the technology is expected to be in trillions of US dollars. This effect is industry dependent. AI also has the potential to create novel industries.

Consumer-based industries being in a position to collect data from individuals have benefitted from artificial intelligence. The data, in combination with artificial intelligence, allows the businesses within the industry to offer personalized service. Organizations through artificial intelligence are now able to personalize promotional messages. Some are now able to customize prices to individual clients. With regard to the personalization of services offered, organizations should take into consideration related laws. Consumer-based industries that have been able to take advantage of AI include those involved in marketing and sales.

They do use artificial intelligence in the management of customer service. Manufacturing industries have also benefitted from the use of artificial intelligence. Here, the use of AI has reduced production costs. Artificial intelligence in manufacturing is being used to improve efficiency. Organizations are using AI to increase the accuracy of their production process. Production is being carried out by businesses at speeds higher than ever before. Manufacturers have now improved their capacity for product manufacture through the use of artificial intelligence. The industry has been able to improve the safety levels of those working within them using the power of AI. Artificial intelligence has also helped in the production of a better quality of products.

In the supply chain industry, the beneficiaries of artificial intelligence have been the businesses that deal with consumer goods. They are experiencing increased accuracy as regards their ability to predict. Artificial intelligence has allowed organizations to focus on drivers of trends. Earlier on predictions had been based on previous outcomes. The effect of AI on supply chain businesses is the reduction of costs related to inventory. The impact is an increase in revenue. There are businesses in the industry that use AI to make decisions.

The decisions made can lead to a reduction in operational costs for such organizations.

Industries dealing with risks like the banking sector are using artificial intelligence to tackle this challenge. They use AI to determine loan underwriting. Artificial intelligence is in use to detect fraud. There is an improvement in performance by the businesses that are using AI. Financial institutions like credit unions use AI to determine how much risk they are willing to take. AI, in this case, helps businesses to make decisions based on the data they have. Artificial intelligence can help organizations in coming up with a risk strategy. AI also helps organizations improve their efficiency regarding the costs of operation.

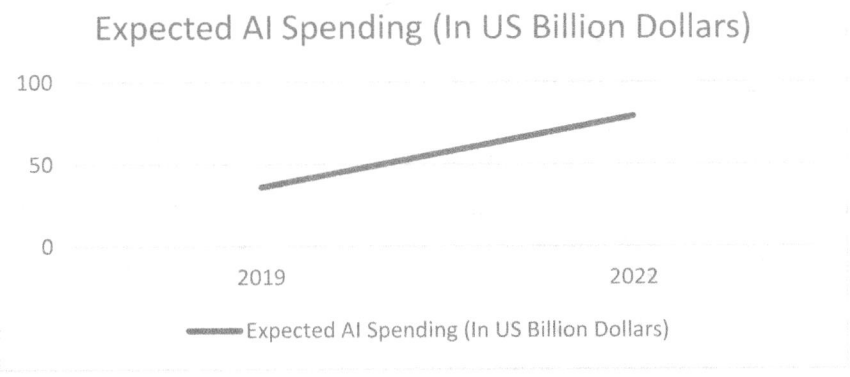

Expected AI Spending (In US Billion Dollars)

Within the service operation industry, artificial intelligence is for operating specific tasks as opposed to its use in a multi-dimensional setting. The actions can include predicting when servers are likely to crash. When determining the artificial intelligence tool to use in organizations, one should think of the purpose of it. Doing so will, over time, lead to shifting the thought process of such tools from the operational context to one of design. The design element should consider the ability of the AI tool to continuously collect data for it to be effective in the industry.

In the field of product development, the design aspects are benefiting from artificial intelligence. The design of the products can be of new items or as an improvement of those already in existence. AI is involved in the automation of the testing process of the products. The role of artificial intelligence in product development has its basis on data. Businesses that choose AI to develop products have been able to break down barriers to production. With all the advancements in artificial intelligence, the uptake of the technology in product development is still low. The role of ethics in product development is a hurdle.

The strategy-based industry is using artificial intelligence to expand their businesses. AI use in such scenarios depends on an organization. Different companies utilize AI in varied ways to achieve their overall strategy. The underlying tool that artificial intelligence uses in the industry is data. Countries are coming up with AI-based strategies with objectives driving the direction of the use of AI in the field. Some businesses input artificial intelligence in their overall strategic plans as a tool in the promotion of their organizational strategies. The plans identify artificially intelligent assets. The result is the production of products and services that are intelligent.

Typology of Artificial Intelligence Applications

The classification of artificial intelligence applications is in two ways. One method focuses on how well an application can replicate humans. The classes of applications in this method include reactive machines, those with limited memory, those under mind theory, and those that are self-aware. The alternative classification method focuses on the intelligence capabilities of the applications. The classes here are ANI (Artificial Narrow Intelligence), AGI (Artificial General Intelligence), and ASI (Artificial Super Intelligence).

In the former classification method, the older models are less capable. The latter method showcases artificial intelligence capabilities that can cut across the ages of the AI models.

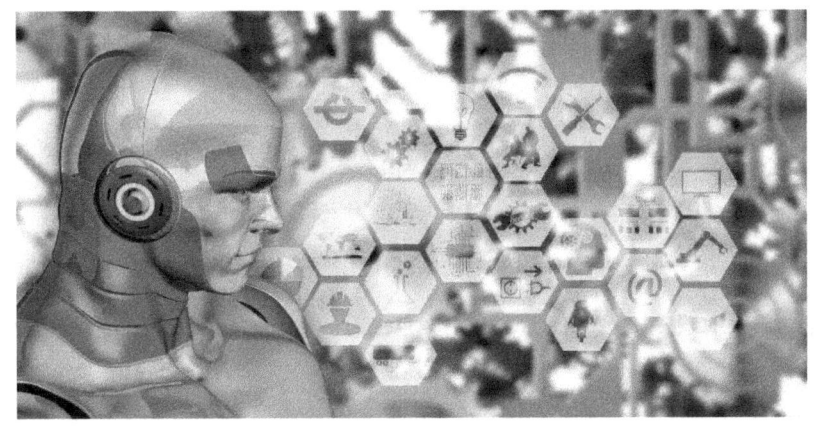

Reactive machines are those that lack memory. These are the oldest type of AI machines. They only respond based on stimuli, which makes them limited in capability. The stimulus they react to can be varied. These machines replicate humans in how they respond. Their lack of memory causes them to be non-learning. The improvement witnessed in these AI machines is not memory-based. The stimulus they respond to can only be of limited combinations. The ability to replicate human responses in previous years within the gaming industry is an example of how the machines work. Limited memory machines, as indicated by their name, have memory capabilities. These capabilities, though, are limited in nature. The capacity gives them the potential to learn. The learning leads to an improvement in response.

The functioning of these machines is memory-based, which allows them to be reactive. Here, the memory in use bases itself on historical data. The data previously collected undergoes conversion into a reference model. The model becomes the basis for decision making by the artificial intelligence machines. Deep-learning devices are examples of AI tools that belong to this category.

The other category in the first way of classification is mind theory machines. These are still in the concept stage or are a work in progress. The devices are expected to have the ability to understand and to identify needs exhibited by other intelligent models. Various companies are working to innovate machines that fall into this category. The overall goal of innovation in this category is to create devices that have a better understanding than humans. The pinnacle in innovation in this class will be to come up with machines that have their belief systems. Some scientists believe that devices will be able to show emotion.

The self-aware category is currently at a stage of hypothesis within the artificial intelligence industry. The belief is that the class will come into existence in the coming decades, if not centuries.

This category personifies the ultimate goal of AI innovation. The aim is to come up with machines that replicate the ability of humans to be self-aware. Also, they will be able to exhibit human levels of intelligence if not superior. The devices are expected to not only understand emotions but also to evoke them. The aim is to create machines that can have independent desires.

ANI (Artificial Narrow Intelligence) machines belong to the alternate classification of artificial intelligence. Such devices currently exist. The complexity showcased by these machines is varied, ranging from simple ones to those that are complex. ANI devices have human-like capabilities. They exhibit autonomy as they carry out specific tasks. The capabilities of the AI devices categorized under ANI are dependent on programming. The dependence causes their competencies to be narrow. With regard to the former classification, ANI encompasses both reactive and limited memory machines. Deep learning and machine learning are part of the ANI capabilities. Some refer to ANI as weak AI.

AGI (Artificial General Intelligence) devices show learning abilities. The machines, unlike ANI devices, can perceive. They exhibit understanding capabilities.

Their functionality levels will be equal to that of humans and can act independently. They can have multiple competencies through the formation of connections across varied domains. These machines will have the added value of reducing training time. Such artificial intelligence does not currently exist but is a goal of AI. The focus of AGI is to replicate human cognitive abilities in machines. Artificial General Intelligence is also known as strong AI. Strong AI based on cognitive capabilities will be able to find solutions to tasks that are not familiar to AGI software.

The achievement of Artificial Super Intelligence is the pinnacle of AI. ASI is a future goal where AI software will be better than humans. Their memory will be more than that of humans. The software will process data faster than humans. The speed of data processing will mean faster decision making. Some believe that such software will threaten the existence of humanity. The software will threaten human culture. Super intelligent software will not suffer the limitation of the cognitive part of humans. This level of innovation will be the height of human creation.

Examples of Artificial Intelligence

Artificial intelligence is in use all around us, yet we may not notice it. Our lives are affected by AI, whether we are aware of it or not. We may interact with AI machines through voice options and those that do not depend on voice. Various industries are incorporating AI into their businesses. Organizations of different sizes are using the software. The goal of AI use is to improve the quality of the lives of humans. The future will have artificial intelligence ingrained in more facets of life.

Business management is running through AI. Organizations are using AI as a competitive edge in the marketplace. Automatic responders work on AI software with process automation, as seen with telecommunication companies and email responses. The responses can be said to act as virtual assistants. The automatic replies reduce associated costs for organizations. Some companies manage their customer service using AI automated responses. Organizations can forecast using AI software whose prediction as it does not rely on previous inputs but drivers of processes. In this manner, businesses get better management. Customers can receive support online from institutions through AI.

Within the e-commerce industry, AI drives the use of smart searches. As one searches for an item, suggestions come up that have their basis as AI. Depending on the level of AI, businesses can personalize experiences for customers of their e-commerce platforms. An organization can use AI as a tool of prediction on expected purchases from potential clients. For e-commerce, the ability to detect and prevent fraud is a role charged to AI. While taking into account related regulations, businesses are using AI to achieve dynamic price enhancement. These AI tools are assisting companies in improving their sales levels. E-commerce is in use in organizations of different sizes.

The marketing industry is one that is incorporating AI in a number of its processes. It is using AI to make recommendations to potential clients, as seen with the use of beacons that transmit recommended offers to people as they pass near brick and mortar stores. Within social media, the industry is using AI to market on the newsfeed of those on the platforms. AI is in use by marketers for ad targeting. There are AI tools that can segment customers, therefore, allowing for ad optimization.

Marketing companies are sifting through customer sentiments on their promotions via the use of AI.

With the advent of global security challenges, companies within the security industry are turning to AI for surveillance purposes. The world is witnessing the use of drones that have a basis on artificial intelligence to target areas of security risk. Telecommunication companies are using voice recognition AI-driven tools to assist clients in protecting their communication devices. We now have phones are requiring or offering fingerprint AI technology as a security feature. Companies are producing machines that can scan the eye as a security feature. AI dependent software is in development to protect against attacks in cyberspace.

The transport industry is using AI software in the form of drones that are reaching far-flung areas in record time. There are countries using drones to transport products that are for urgent use, for example, pharmaceuticals. The industry, in combination with security, is sending vehicles that are not controlled by humans. Some cities are using artificial intelligence to manage the movement of cars. Applications like Uber and Lyft are examples of how AI is causing a change in the transport industry.

Self-driven vehicles are now in use and are under further development by various companies.

The human resource industry is not remaining behind in using AI. The industry is at the receiving end of AI development as some feel the tools developed are taking the place of humans. The field is at a crossroads as regards using AI. On the one hand, it is increasing efficiency while on the other hand, it may mean laying off workers. There is software that is being used to sift through the resumes that companies are receiving. AI software is now available that human resource managers can use to facilitate mentoring. Some companies are using the tools as a provision of continual learning platforms.

Healthcare companies are reaping benefits from the field of artificial intelligence. The area of pathology is seeing AI-based tools that are assisting in the reading of biopsies. When linked with administrative components, healthcare companies are using AI to help medical workers provide individualized care to patients. There are surgical robots that some hospitals are incorporating in management regimens. Some are showing higher levels of accuracy in comparison to humans.

The focus in the healthcare industry is to provide more time for medical workers to interact with patients as opposed to working on administrative issues. The pharmaceutical industry is using AI software to come up with newer and better medicines.

Chapter 2: Artificial Intelligence: Promises and Challenges

As with all innovation, artificial intelligence comes with its challenges and opportunities. One comes across these aspects of artificial intelligence as they interact with the software in various forms in the goings-on of daily life. Individuals and organizations can choose to take advantage of the opportunities that artificial intelligence presents. The challenges should not be a reason to discourage the use of artificial intelligence. As with other human innovations, the focus instead should be on continuously improving the same. The occurrence of the opportunities and challenges can be due to the one interacting with them.

The promises and challenges that artificial intelligence is presenting are sometimes industry-specific. The artificial intelligence software may work well in one industry as opposed to another. The occurrence may be due to the kind of input that a specific field gives to the artificial intelligence software. What may be an opportunity in one industry may turn out to be a challenge in another. The focus here should be on whether the promise or challenge cuts across industries or is limited to a particular field. Customer-centric industries may, for example, be able to get access to more data than those that are not.

Artificial intelligence, though considered as being independent of machines, can be affected by the human element. For the artificial intelligence software that relies on input from humanity, the biasedness within individuals may affect how it works. The artificial intelligence software that relies on historical data assumes that the data earlier present is accurate. The effect of the human element can present as a challenge. Artificial intelligence software that is dependent on programming by people has its limitations as being defined by those who created it. The knowledge humans have limits or create opportunities in artificial intelligence.

The field of artificial intelligence is one that embraces innovation, yet this characteristic can be a source of a challenge as it is of opportunity. There are scenarios where improvements in historical artificial intelligence software lead to unprecedented development. The changes can cause unexpected negative occurrences. Innovation in planes is, in some cases, leading to crashes of airplanes. Though new artificial intelligence software undergoes tests before releasing into the marketplace, some effects present when used on a massive scale.

Some innovations do get new uses once released into the general population.

Artificial intelligence software concerning both promises and challenges has to contend with the political aspect of human life. There are innovations within the artificial intelligence space that do not attract political goodwill. These innovations are being shut down from a global perspective, with these occurrences presenting a challenge to those interested in developing artificial intelligence software. Political goodwill can be a source of opportunity for those working on artificial intelligence software seen to be helpful to society. Countries are using artificial intelligence to determine their stand globally on an economic scale.

Promises

The opportunities artificial intelligence portends are massive. As we interact with AI software in our daily lives, we realize the benefits they bring. The benefits cut across various aspects of our lives. Some AI tools are so much a part of our lives that ignoring them would be to our detriment. The benefits are changing our world as we know it. Those who are undermining the potential of artificial intelligence are doing a disservice to themselves. There is an expectation that in the future, artificial intelligence software will improve on its intelligence aspect.

Socially, AI is proving to be full of promise. The use of AI is supporting sustainability, for example, in the use of software that is creating social communication channels that are low in cost. The reduction in cost creates easily sustainable communication models that are helping in building the social fabric of humanity. These AI-based channels are acting as a medium through, which, various cultures are interacting, allowing for an appreciation of vast human social structures. Social media systems based on AI are changing the way humans are interacting with one another. Some contend that the channels are creating a basis for the realization of a global culture.

The promise of AI is affecting the political or governance aspects of human life positively. Candidates who previously had the right ideology but were limited by, for example, the cost of reaching their intended audience can now ride on AI channels. Such channels are providing affordable ways of reaching out to populations. There are cases where their use has led to political upsets that are considered to be positive. Some individuals in the political space are using AI to solve social concerns.

Those passionate about the environment are reaping the fruits of AI. AI-backed communication systems are reducing the need to use paper. Using AI in this manner is saving trees. The field of meteorology is using AI to predict weather patterns. The prediction is allowing for preparation in case of weather forecasts that are heading in the wrong direction. Such data from AI systems are in use for saving lives. Transport companies are relying on the weather warnings to determine how safe it is to embark on travel. There are global standards of weather patterns that airlines can fly in.

AI is showing promise in the education sector with learning accessibility rising due to its use. E-learning is an example of how AI is changing how humanity is educating itself. It is creating opportunities for learning in sectors that are becoming available due to the advent of AI. The possibilities include the fields of data research and data mining. Global standards on education are increasingly becoming popular as AI is connecting the world. There is an increase in the number of individuals receiving education powered by AI software. AI is making education affordable and convenient.

Benefits for Business

The benefits that businesses are deriving from AI is immense. From machine learning software to deep learning tools, the opportunities the field of AI is presenting to the world of business is vast. Organizations are using AI to build a competitive edge within the marketplace. The importance is visible in highly competitive industries. For companies to benefit from AI, they have to invest in data mining software as data drives AI. Latter AI software combines data with algorithms to produce analytical models that businesses are using to make decisions.

In highly competitive industries, efficiency is improved using AI. The automation of processes allows for better customer experience for business clients. The automation may include AI software that can give insight on ways to improve operations. The knowledge from the comprehension assists companies to work on their operational weaknesses. Cloud computing, as an example, uses AI to negate the disadvantages of servers being present on business sites. There is AI software that helps businesses reduce their costs of operation. AI is helping organizations operationally by reducing human workload, therefore, allowing people to focus on core business operations.

The sales and marketing aspects of businesses are transforming by the advent of AI use. Individualized customer marketing is, now a reality. Targeted ads on websites and search engines are now part and parcel of today's marketing program for businesses. On the sales front, AI-backed CRM (Customer Relationship Management) software is commonplace as concerns working the sales process. These tools allow for aggregation of large amounts of data giving insights that were not previously available. Companies are using AI-based beacons to personalize shopping experiences from pricing to offers. Marketing ads are now a staple on social media platforms.

The human resource aspect of businesses is benefitting from AI in a variety of ways. The development of AI is creating opportunities for businesses to access the global workforce at the touch of a button. Organizations are seeing a surge in the availability of freelance workers willing to work remotely for industries able to take advantage of the opportunity. The ability to work remotely can be a competitive edge for businesses and can be a way to reduce human resource costs, given the varied pay structures on a global scale.

Businesses are transforming how they interact with customers using AI-backed systems. Some companies have incorporated AI tools like chatbots as part of their customer care strategy. Studies are showing that the use of automated customer care tools is improving levels of customer engagement. Moreover, it can augment the human aspect of customer service with people taking over from the AI systems when the task is beyond the abilities of the AI software. AI systems are in use for follow up purposes on customers by businesses for cases that require feedback by the company to the client.

Benefits for Economies

Economies with a global outlook are now using AI to advance their interests in the world stage. Some are choosing to be the hub of specialization in the field of AI or AI-backed industries. Countries are investing in AI-based military hardware as a way of protecting their economic interests. For economies, the marketplace has increased as the world connects via AI-backed systems. Countries can trade conveniently at the touch of a button. Increased competition between economies is leading to better product quality for customers. Wealth generation is happening at a rate never seen before through the use of AI.

Economies are investing in IT systems that are backed by AI to improve the efficiency with which they are generating wealth. Industries that are supporting AI are springing up changing economies worldwide. Global wealth is favoring economies embracing the use of AI. Countries are making economic decisions from data collected through AI-backed IT software. The use of AI in the IT industry is giving a chance to economies that are lacking traditional resources to generate wealth based on human resource AI skills. Economies are leveraging on this opportunity that AI is presenting.

Virtual economies are now a reality with AI playing a role in their growth. AI-backed systems are introducing digital currencies as an alternative to traditional monies in economies. Though still being worked on, there are those expecting the future economy to be cashless. AI systems like blockchain are expected to be the backbone of the digital currencies. Social media companies are looking to take advantage of these developments. Economies, where the companies are based, are growing at an unprecedented rate. Virtual jobs are now a part of any economy that is growing.

Some economies are transforming by using AI-backed manufacturing processes which are leading to a shift in where companies are choosing to base their production facilities. Economies are growing in size by investing in AI-focused systems, leading to a reduction in costs of manufacture, therefore, being competitive in the global marketplace. Cost reduction is in both labor and capital aspects. Industries incorporating AI into their manufacturing processes are seeing positive changes on a large scale. Manufacturers are increasing their efficiency by using AI automated systems. Countries are changing the source of their economic strength by investing in AI-backed manufacturing industries.

AI is proving to be an integral part of the healthcare system with its use in the diagnostic stage of treatment.

The healthcare segment of clinical research is benefitting from AI-backed systems with economies transforming their workforce with interventions generated from such investigations. Some countries that were losing their workforce at a younger age are using AI to improve the length of life. Economies are growing by generating new ways of treating diseases, transforming some countries into medical hubs. As people travel to these economies for healthcare, they pump money into the sectors.

Contribution to Productivity, Growth, and Innovation

AI is contributing to the growth, innovation, and productivity of economies and businesses. The AI model in use is the one determining the extent of the contribution. The effect of AI on growth, innovation, and productivity is expected to increase as AI incorporation increases. Businesses and economies should be aware of the factors that could slow down this contribution and take steps to minimize them. It is best to consider AI as an augmenting factor to labor and capital, which are traditionally factors increasing productivity, growth, and innovation.

In terms of growth, AI is contributing to economies with some showing accelerated growth levels. Some countries are changing their economic prospects by working on becoming a hub, for example, for manufacturing by using AI to reduce their costs of production. The result is they are increasing their global competitiveness. Businesses are increasing their capacities by investing in AI tools that are increasing their efficiency levels. The economies investing in AI tools are witnessing a growth of opportunities as their capacity increases. Businesses and economies are seeing a growth of jobs associated with AI.

The contribution of AI to innovation is undisputable within the field of healthcare. Here, AI development is augmenting humans as they seek better healthcare outcomes. From diagnostic tools supported by AI to those assisting in surgical procedures, AI is driving innovation in the medical arena. Innovation in Information Technology is increasing as the demand for improvement in AI is being pursued. Countries are innovating new ways of improving their military capability using the power of AI, for example, by using drones instead of humans. Industrial growth due to AI is now a reality as businesses supporting AI are increasing.

AI is contributing to productivity by increasing efficiency levels. The speed at which processes can be carried out using AI is at times higher than that of humans. AI can carry out tasks with increased accuracy even at higher speeds, through automation. The combination of AI abilities is leading to increased output levels. There is an increase in production volumes in economies and businesses that are investing in AI. AI augmenting humans in work is leading to increased productivity levels as humans are focusing on core tasks. Individuals experience freedom from mundane tasks as they can focus on those requiring creativity.

Businesses and economies can ignore AI to their detriment as it is becoming a part of daily life. Entities using AI are expected to have a competitive edge in the market as consumers get a better experience from them. In a global marketplace, AI can help organizations reach far-flung regions at a minimal cost. Businesses can use AI to improve service or product offerings. The effectiveness of AI for organizations will depend on its use. Capital owners are at an advantage concerning benefitting from AI.

Challenges

No human invention is perfect, and AI is no exception. Businesses and economies should be aware of the issues associated with AI and put into place systems that negate the hurdles or turn them into strengths. It is better to face the negatives than ignore them and suffer down the line. Being aware of the challenges puts businesses and economies in a better position to navigate the world of AI. The awareness may become a driver of future innovation through a goal of improving on weaknesses of AI. Acceptance of the issues allows entities to know their limitations in advance.

Businesses are at risk of losing customer loyalty due to the use of AI if clients feel interaction with an entity lacks a human element. The social aspect of human interactions is critical. Potential clients may feel a business is choosing machines over humans, for example, if jobs are lost. The message an organization may portray by incorporating AI may be ill-received by the public. Customers may feel profit is the focus of a business as opposed to humans. The lack of cognitive ability by AI tools can portray a business as inhuman.

Studies from various quarters portend that as AI becomes part of daily life, the result will be job losses. The reasoning is that AI has higher capabilities than humans in terms of, for example, speed and accuracy. Loss of jobs would lead to a breakdown in the social fabric as individuals look for means to survive. Stress levels are expected to increase with security challenges expanding. Individual levels of stress will increase as they look to adapt to changes in the marketplace. The reduced levels of human interaction are expected to increase the negative emotions experienced.

The growth of fear is a challenge associated with the advent of AI. The emotion is leading to a reduction in productivity as individuals are battling with the idea that they could be redundant shortly as machines are looking to take over their jobs. Others are having to contend with changes in their job descriptions to keep their work opportunities alive. Families are facing changes in job locations to stay in employment as companies are changing business strategies, for example, by changing manufacturing locations. Businesses are taking these actions in a bid to remain competitive.

AI challenges include building trust, AI-human interface concerns, Information Technology infrastructure, and investments. These challenges affect how the general population is adapting to AI in their daily lives. The issues need addressing so that humanity can take advantage of the full potential of AI. Each of the challenging elements may require a different approach. The overall focus should be on how to turn the challenges into opportunities. The process would allow for continuous improvement within the field of AI. The issues will affect how economies choose to strategize for future generations.

Building Trust

Given the emotive issues raised concerning the effect of AI on the general population, building trust should be part of the agenda of future AI strategy. Issues of ethics, data protection, safety, and transparency are critical to consider. The approach to these concerns determines whether the trust is built or lost. Without confidence, the AI journey will attract more challenges from a variety of quarters, including the political class. The more the challenges, the slower the progress to full AI potential will be. Trust will make the transition to AI smoother.

An element that the AI industry is tackling to build trust is ethics. The hurdle is in creating agreeable boundaries that are acceptable to all stakeholders. Companies are updating and sharing privacy policies with potential customers. The goal of AI in the context of ethics is to come up with an acceptable global standard applicable worldwide. Such an achievement will standardize expectations from customers. Ethics, when appropriately applied, should not strangle AI but instead, promote it. Changes are coming up in ethical standards as issues are arising as AI is undergoing implementation on a much larger scale than ever before.

Data protection is an element of trust-building when it comes to AI. Individuals want to know that their data is safe when they share the same. There have been cases where hackers have gained access to data from companies without authorization. Such occurrences are breaking down trust towards the use of AI in daily life. AI tools dealing with sensitive areas like finance cannot afford to have laxity when it comes to data protection issues. The use of AI tools in the financial industry is facing data protection as a hurdle to building trust in potential clients.

The safety of individuals is an issue when looking at building trust in AI. There are AI tools that are in use as security measures. The challenge is in the boundary between using AI as security without creating the feeling of loss of privacy. When one senses that they are losing their privacy, they tend to shy away from using AI tools. AI tools should not be a source of a security risk as witnessed when an individual has their personal information shared due to a security breach.

Businesses and economies need to embrace transparency to build trust with potential customers. Customers should not get a sense that they are not getting the full information on how their data is in use. There are cases of organizations getting exposed through cases where their affiliates are using the information they are sharing without customer approval. Entities need to share information and update if needed, on how they are using shared data. When potential customers consider an organization as transparent in their dealings, they tend to trust the entities. Organizations should aim to create the right balance of honesty in the eyes of the public.

Artificial Intelligence Human Interface

Humans possess the need to interact with other individuals regularly. This innate need is presenting a challenge to the AI field, for example, for AI tools that do not have a human interface. Some customers may feel offended at the lack of interaction with other humans when transacting with a business. Some report feeling as though they are just a number to the organization. In a world where competition is increasing, entities need to tackle this apparent business challenge.

Artificial Intelligence Human Interface is facing a challenge in the context of augmentation. There are AI tools, for example, in healthcare which are in use to assist in treatment processes. When something goes wrong during treatment, where does the blame lie? At what point is the responsibility of the individual and at what point is it a machine issue? Businesses that are incorporating AI tools in areas like customer service are facing the challenge of determining when to shift a conversation to include the human element. If not done correctly, incorporating AI tools may become a weakness for a business.

The advantages of AI tools are not changing the way some individuals perceive the software. Some consider organizations that revert to AI tools without a human interface as untrustworthy. They are those who believe that virtual assistants are not knowledgeable about products and services as compared to individuals on location. The perception is presenting a challenge to organizations looking to scale using AI-backed services. Context is a factor that determines the quality of customer service and can be compromised. Some AI tools have their programming done in a limited manner, for example, in the variety or specificity of choices the software presents.

Currently, AI tools do not have cognitive abilities, which are a factor in the challenges facing the AI-Human interface. AI tools programming is generally focusing on dealing with specific tasks. When an AI tool is facing an option or input that is not part of its program, it gets challenged. The balance between taking advantage of the full potential of an AI tool and knowing when to introduce a human interface is a challenge for organizations. Businesses must determine what is delegatable to AI tools and where individuals will add value.

Speed can be an element of the AI-Human interface challenge, for example, when an individual is in a faraway location from the AI tool in use. There is the element of time in terms of how fast an individual can respond to a query forwarded through AI tools. The person on the other end may assume that they are in contact with a human and not a machine. Time zone differences may lead to a slowing down of responses. A business may receive negative reviews on their services due to this challenge.

Insufficient IT Infrastructure

The challenge of IT infrastructure insufficiency is one that can be solved. The underlying focus in dealing with this challenge is to find out where the source of lack is. It can be an issue of processing power, hardware, skill, and even support. Each of these sources will require different responses. Businesses should opt for solutions that will have the most impact in the case of competing interests for limited resources meant to deal with the concerns. Organizations should also consider the most affordable solutions. Approaching the challenges in such a manner may help a business sort out more than one source of the lack at any given time.

Given the large volumes of data associated with better models of AI tools, businesses that choose to invest in the latest AI devices should prepare to have higher levels of processing power for the data. Within the AI strategy of an organization, an entity may choose to focus on improving the processing power of its AI tools over time. The advent of cloud computing is proving to be a solution to this challenge. Some companies are utilizing the power of parallel processors.

IT infrastructure can lack in terms of hardware. AI that is dependent on machine learning may require a company to have its machines at location. With AI being a dynamic field, what may be of use today may be unusable shortly. Companies may need to keep updating their AI tools, which may come at a cost that is unsustainable to the organization. Businesses may consider how long an AI tool they are to purchase will remain relevant. The versatility of such devices can help a company keep its costs low.

Without the right skill, the use of AI will be a challenge for businesses. Some AI-based devices are facing a limited market globally due to a lack of supportive skills needed for them to work effectively. Some companies are choosing to limit where their items are on sale to deal with this challenge. The available workforce may prove expensive to some organizations, which translates to investing in some AI tools untenable. The gains from AI may be slowed down for organizations that are relying on skilled labor that is overburdened. In such cases, the investment in AI tools may prove to be a costly mistake.

AI as a discipline is dependent on other skills for it to be effective in its goals. Lack of support from industries that complement IT infrastructure is proving to be a challenge for some organizations. An example is the lack of a reliable internet connection that may be a requirement for some AI tools to function. Several regions globally still lack, for example, fiber connections that are a requirement for processing high data volumes, which is a feature of newer models of AI tools. AI devices relying on other sources of data may work incorrectly when the providers of such information are missing.

Inadequate Investment for Implementation

Organizations should withhold investing in AI tools if the investment is inadequate for useful implementation. Businesses should consider all the elements required for AI implementation before starting the process. Some facets to consider for efficient AI implementation include the availability of human skill, hardware, supportive industries, and goodwill. An organization choosing to ignore any of the facets may end up with an incomplete project which may add unnecessary costs to the business.

For any AI tool, a business is looking to incorporate in its processes, the human skill element is critical. Companies should tackle this aspect in advance. If applicable, the level of knowledge and skill of staff needs confirmation before investing in AI tools. A plan on acquiring the skills can be part of the implementation strategy. Some companies are establishing modules for continuous training to keep the skill levels of staff up to date. Such a program is to ensure that the full potential of the AI investment is in reach. Organizations can use the power of Frequently Asked Questions (FAQs) to assist staff in handling challenges they are facing during AI implementation.

An efficient AI strategy takes into account all the components of hardware required to implement AI successfully. During the planning phase, an organization should consider if all the hardware elements are easily accessible. Scenarios, where an AI tool is not functioning because an organization is waiting for the shipping of a piece of equipment, is common. Businesses end up losing opportunities due to such occurrences. Adequate investment in this context may mean having a spare piece of hardware at any time.

The successful implementation of AI can be dependent on supportive services and products. Businesses looking to implement AI should ensure such services exist and are readily available. It may be prudent to hold off instituting AI if there is no guarantee of support for the same. Going ahead with no backing may present a business challenge in the future. Lack of fiber installation can be why an organization cannot implement AI in some regions. The effect is the slowing down of benefits connected to such AI opportunities from reaching the firm.

Several organizations may forget to consider the political aspect of AI implementation. Companies need to invest in AI in regions where there is political goodwill for the same. A lack of political goodwill for AI projects can cost businesses. Politics can clash with AI implementation as people may feel that the incorporation of AI will take away their jobs. A prudent organization will look at the impact of the political climate concerning AI implementation before deciding to invest. Finding a balance between investing for the future and taking care of political interests can be the key for a company to survive in the marketplace.

Chapter 3: How Artificial Intelligence is Changing Business Processes

Business process as we know it is changing, thanks to the power of artificial intelligence. The rate at which AI is changing how business is processing information is dizzying. More organizations are taking advantage of the promises of artificial intelligence. The difference AI is making in such organizations is clear.

Companies are increasing their levels of efficiency to never seen before standards using AI tools. Operations that were taking days to complete are now over in hours or less. Some AI tools can achieve speed without losing quality. Workers are getting much-needed support from AI tools leading to better outcomes.

The more the AI tools get incorporated into business processes, the more innovations are increasing. Modifications of AI tools are occurring as their use is growing in businesses. The workforce is getting more creative to survive the onslaught by machines. They are also getting more time to create, as AI devices are taking over mundane tasks.

Artificial intelligence has brought the facet of convenience to business processes. Procedures previously taking hours to complete can now be finished in comparatively shorter periods, sometimes with the click of a button. Industries based on this factor brought in by AI are blossoming, for example, purely e-commerce based businesses.

Businesses are now able to take advantage of growth opportunities that were not available before the advent of AI. Brick and mortar stores can run e-commerce shops in tandem. Businesses can reach customers that were previously unavailable to them at a minimal cost. The range of products companies can offer customers are also growing.

AI is converting manual business processes into automated procedures. Companies are using this opportunity to stay ahead of the competition as automation is allowing them to reach more customers in shorter periods. Faster processing is attracting more clients to businesses. Automation is also lowering business costs.

Businesses are now able to incorporate segmentation into their internal and external processes. The effect on marketing is companies having the ability to personalize product offerings. The use of beacons as a way of optimizing sales promotions is becoming more common among business marketing strategies.

Businesses are adding value to products and services using AI-backed tools. Some products previously unavailable in digital forms are now in production in eformats thanks to AI-driven possibilities. Examples include producing reading materials in digital formats as well as traditional hardcover types. The overall effect is increasing value for businesses in terms of sales opportunities.

The productivity from the AI-backed business process is growing faster than traditional models. Some tasks are processing at faster rates as machines are taking over mundane tasks. Increasing productivity is leading to cost savings for companies. Some AI-backed devices are allowing companies to function for longer hours.

The creation of new jobs is occurring as AI is changing business processes. Jobs are paying more in some scenarios. Opportunities involving less mundane tasks are characterizing the AI-backed changes in business processes. The new opportunities require individuals to change their way of working. Cross-functional skills are a requirement in the jobs.

AI is changing business processes including, customer acquisition, new customer service types, personal assistants in businesses, marketing research, human resource and hiring, and the sales process. The extent to which the methods are changing is proportional to the investment businesses are putting in concerning AI-backed tools.

New Customer Acquisition Process

The cost of acquiring a new customer for businesses is an element of company processes that adds to the cost of doing business. Getting a new client is more expensive than retaining one to companies. Regardless of the negatives, organizations must keep the process of acquiring customers active to avoid becoming redundant in the marketplace. Businesses are spending huge volumes keeping themselves in the mind of potential clients. Budgets for such activities are consistently in renewal as organizations are trying to maintain their competitive edge. AI is changing the way companies are approaching the process of expanding their customer base.

The advent of AI-backed e-commerce opportunities is providing new ways for businesses to acquire customers. E-commerce platforms are adding a convenience factor that brick and mortar concepts are unable to present to the market. Some organizations are choosing to combine both ideas for greater appeal. Some organizations attract customers looking to purchase online without visiting stores physically. Businesses are offering products and services to attract such customers, for example, free shipping. Customers can sign up for membership online to stores with the click of a button. Businesses are gaining customers who have never stepped into their stores physically.

AI through social media is giving businesses new tools with which they can reach new customers. From business pages to targeting ads, organizations can use varied ways to acquire new clients. The social media platforms are providing tools that allow personalization of business pages. Organizations can share information that is attractive to potential clients through such media. The modern customer is expecting that businesses have a social media site that they can reach them through if need be. Potential customers are raising queries through the social media news feed and expecting a rapid response from organizations. The effectiveness of social media platforms is no longer debatable.

AI tools are providing opportunities for businesses to target with more accuracy potential customers. The effect is opportunities to personalize product and service offerings to new customers. Sites like Facebook and Instagram, are allowing for targeting based on varied criteria include geography, age, and interests. Businesses can work on the timing of their acquisition messages for effectiveness. Since the targeting tools are using data and AI, the accuracy levels are making it easier for businesses to acquire new customers faster. Organizations are, therefore, spending less money on expanding their client base. AI is allowing companies to place targeted ads on websites that are appealing to the kind of clients they want to acquire.

In the world of marketing, AI has created business opportunities to expand the market for businesses through the use of beacons. These devices can, through the help of AI, interact with potential customers passing near a brick and mortar store. The tools will send messages that are personalized by businesses encouraging customers to walk into the stores. Organizations are taking advantage of these tools to send individualized messages while taking advantage of proximity to the potential client as a means of acquiring new clients.

The power of AI-backed search engines is a reality for the majority of the world's population. The company Google is one of the largest in the world, with the largest search engine. Businesses are using the platforms to reach out to new customers by aligning their websites and contents to be the digital place of choice for clients. Organizations are investing in keywords that are optimizing their sites for search engines to rank them higher than the competition. An industry based on search engine optimization is budding based on the opportunities the AI-backed systems are providing to businesses.

AI-backed software is allowing businesses to acquire new customers on a global scale. Industries based on digital products and services can transact across borders. AI-backed payment systems are making the process easier, effectively expanding the potential client base for businesses that are embracing the tools. AI-backed logistic tools are giving customers, real-time ability to track shipments, increasing trust levels between companies, and customers leading to the increasing willingness of clients to take risks concerning purchasing from faraway locations. The convenience associated with digital products shipping instantly via AI-backed tools is increasing the global reach for new customers for businesses.

New Type of Customer Services

AI is creating new types of customer services by providing opportunities within the marketplace that were unavailable in yesteryears. These services are gaining ground as they are less tedious than traditional services. The services do not have to contend with the challenges of geographical borders. The services are coming up as a result of innovation driven by AI. Some regions require less regulation, which is encouraging the fast growth rate witnessed in this sector. Some of the services are springing up in support of AI-backed systems. Such services include those that make sense only if backed by AI tools.

The digital payment industry is a service that is growing based on the premise of AI. Businesses relying on digital products are requiring means of payment. Companies working as escrows on online platforms give the much-needed trust factor for online transactions. The companies provide a medium through which both the buyer and the seller can get protection from fraudulent activities. These payment modes are also providing convenience as one can release or receive money at the click of a button. Without these services, businesses transacting online would be less efficient. The convenience factor the companies are having over brick and mortar stores would also reduce considerably.

Education traditionally is held in a physical location where students gather to listen to knowledge. AI is creating platforms that are in use in online classrooms. The advantages of this mode of education include increasing the global marketplace for education. The diversity allows for increased multi-cultural understanding, which may lead to a more unified global society. The convenience element that this service offers cannot be understated. Online educational platforms are increasing access to education, which can be a media to achieve equality. Some platforms are giving options for potential students to personalize their educational journey.

Digital TV service is running on AI-backed systems, and its growth is proportional to the availability of AI in different regions. This service is pushing out traditional forms of entertainment with its convenience factor. Some providers are allowing for more capacity per payment made with clients able to stream services on more than one device. Users can access providers on a global scale without limitations associated with traditional boundaries as providers can present offerings worldwide. Competition within this service is stiff with providers changing how they provide entertainment to potential clients. AI support services must be in place for one to take advantage of these services.

The industry of freelancing has its backbone as AI. Individuals are now able to work from remote locations while being as efficient, if not more, as being in a physical office. AI is allowing economies to work on a 24-hour basis through freelancing. Those from across the globe can work while others are resting using AI-backed tools. Teamwork is possible using AI without the hurdle of geographical boundaries. Companies have the opportunity to outsource tasks that not core to the business on a global scale. Outsourcing globally could mean saving costs for businesses.

AI is responsible for the growth of online stores with Amazon proving its worth on a global scale. Brick and mortar stores are also able to take advantage of this AI-backed tool, therefore, taking advantage of the online marketplace. Sales are increasing for businesses using this platform. The stores can be set up conveniently with companies now available providing templates for online stores. The updating of store offerings is convenient with AI-backed tools. Dropshipping services through companies like Shopify are now available riding on AI.

Such online stores require minimal investments, therefore, lowering operational costs.

Since AI is running on the web, services that support the web industry are coming up. AI support services like fiber-related companies are becoming commonplace across the globe. Independent businesses that traditionally would require large sums of investments like TV stations are coming up at a fraction of the cost. Such companies are taking advantage of AI-backed organizations like YouTube. Here, they are building channels with some achieving high levels of views from minimal initial investments. Other services include creating digital products that are selling via online platforms. The products include digital games, stickers, and fliers. Graphic designers can now provide services online as opposed to the traditional methods of yesteryears.

Using a Virtual Personal Assistant in Your Business

Traditionally, a virtual personal assistant is an individual who is remotely connecting to the clients of a business. AI is changing the industry at dizzying speed. Machines are taking the place of humans in the field of virtual personal assistants. AI devices acting as personal assistants include chatbots and autoresponders. The devices are programmed to lead one through, for example, a series of questions to get a predetermined solution. Some can communicate in various languages expanding the options available to business customers.

The opportunities that AI is bringing to businesses that are opting to use virtual personal assistants include lowering costs of doing business. An organization using virtual personal assistants does not have to take into consideration labor laws associated with a human interface. Issues of overtime, medical leave, and union rights do not arise. AI-backed virtual personal assistants can work round the clock, unlike their human counterparts. The devices work remotely and require no direct supervision. Businesses can upgrade AI whenever it is feasible, ensuring they are up to speed with the standards in their industry. AI-backed virtual personal assistants can be programmed to deal with specific tasks.

As with all human developments, virtual personal assistants come with their challenges. Unlike humans, virtual personal assistants, even those associated with AI, lack cognitive capabilities. They are unable to connect emotionally with business clients, which can portray a business as having a cold persona. Depending on the industry an organization belongs to, it may be a source of negative feedback from potential clients. Politically there may be pushback on replacing humans with virtual personal assistants as economies struggle with unemployment rates. Getting the right balance between machine and humans, for businesses looking to incorporate AI into their operations, is a common challenge.

A strategic mindset should be in place when businesses are looking to incorporate AI-backed virtual personal assistants into their processes. The incorporation of the tools into the overall business strategic plan gives the best direction to follow. The business policy should allow for counter-checking on which virtual personal assistant would work best for overall organizational strategy. The goal of incorporating virtual personal assistants should be easing the movement of a business towards its end goal. AI-backed virtual personal assistants may provide a competitive strategic edge over other market players.

Companies providing AI-backed virtual personal assistants are many with providers located all over the world. Some virtual personal assistants do not need complex support systems. Some providers allow businesses to use them for free while some charge. Some social media platforms are allowing companies to personalize their own virtual personal assistants at no extra cost to the business. Businesses can create their own virtual personal assistants online like chatbots. Organizations can consider choosing virtual personal assistants that come with the added benefit of machine learning. Companies should work with authentic sources of AI-backed virtual personal assistants to avoid unnecessary downtimes.

When determining the virtual personal assistant to use, a business looks at the context of the industry. There are industries, for example, customer service, that are accepting of virtual personal assistants. Clients are on board with technology and are focused on the convenience factor. Other industries are not so liberal, for example, those where clients are looking for a human touch like healthcare, and counseling. In the latter scenario, potential clients may find it offensive working with virtual personal assistants. Businesses in sensitive industries should get the right balance between the use of virtual personal assistants, and delegation to the human workforce.

There are various industries where using a virtual personal assistant in business would be beneficial. Examples include the customer care industry where potential clients are more focused on finding a solution and hold liberal views when it comes to communication methods. Some service-based industries, like travel agencies, are incorporating virtual personal assistants with success. There are virtual personal assistants now available for personal use. Some businesses within the manufacturing industry are using virtual personal assistants to automate processes as a means of freeing the workforce to focus on their core job description. Companies involved in retail marketing are also using virtual personal assistants.

How Artificial Intelligence Is Changing Marketing Research

Historically, marketing research involves sending the human workforce to chosen areas where they would ask questions to a predetermined sample population. The answers would give a glimpse of the opinion of the market towards a product or service. Data would come from the answers, and analysis would yield a conclusion. The whole process was tedious and would take months, even years, to conclude. The cost implication of the process was high in terms of money and time. The possibility of a research result, being redundant by the time it was submitted, was a possible reality.

Artificial intelligence is changing marketing research by erasing some of the challenges previously experienced within the industry. Marketing research companies can now carry out work conveniently and can, therefore, interact with larger populations. The result is higher levels of accuracy. The convenience is breaking down barriers that had fewer people willing to take part in surveys. For the research done remotely, the privacy factor associated is giving researchers a larger pool of individuals willing to share their opinions on issues. The timing of responses is also now convenient with respondents choosing appropriate times of answering surveys.

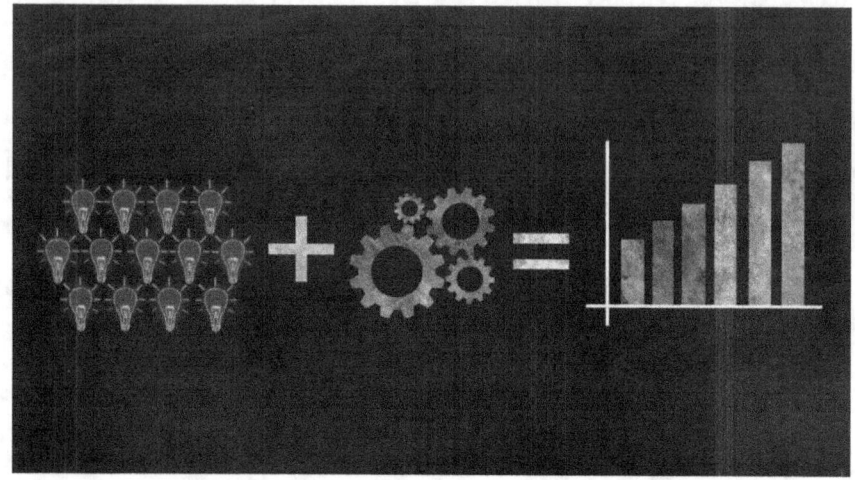

Artificial intelligence is giving marketing research companies opportunities to expand their reach of possible respondents. Through AI, market research is now possible on a global scale. Geographical barriers are a thing of the past for market research companies that are embracing the use of AI in their research processes. AI is tackling barriers like cultures that were a hindrance to marketing research. Examples include those that inhibit part of the population from giving their opinions on varied issues. AI is allowing individuals to share their views without necessarily revealing who they are, which is allowing more people to speak.

With AI, marketing research companies can segment their target population at a faster rate. At the click of a button, a company can determine the boundaries of the sample population. The limits to be defined can include geography, age, sex, gender, religion, and work. The potential of targeting accuracy is more with AI-backed marketing research tools. Targeting with AI gives results that are a better reflection of reality. Marketing research companies working on a global scale can determine the appropriate timing of the process per region. The result may be an increase in the number of respondents per survey that is carried out.

Artificial intelligence is allowing marketing research companies to carry out their processes online. The benefits of online platforms are numerous, including the opportunity for marketing research companies to tweak surveys as results are coming in. Marketing research companies can add or reduce the scope of their research as respondents are acting. There are marketing research companies that are creating applications that send out surveys as needed. A notification would pop up on the devices of potential respondents, giving them a chance to share their opinions conveniently. Online research is allowing companies to see results in real-time.

The advent of AI in marketing research has considerably reduced the turnaround time for results. Marketing research companies can see the results as they are trickling in from respondents. The speed is allowing for versatility with marketing research companies carrying out multiple types of research concurrently. The companies can serve many customers while using their resources efficiently. Businesses looking for quick turnaround times from market research companies are at an advantage. The quick turnaround time is proving beneficial to organizations as they can react to the market faster. Businesses can use AI marketing research as a tool to keep abreast of market changes.

AI-backed marketing research methods are devoid of human bias, which may lead to better results. AI is immune to biases that humans are prone to like gender, race, and religion. AI devices can be programmed to get rid of bias. A lack of bias may give new insights previously unseen. Businesses may end up seeing opportunities that were blocked by human biased results. Unbiased insights can be for companies, a source of competitive advantage, growth opportunities, or new markets. AI-backed marketing research that bases itself on machine learning will lessen the bias further. Such results will give real-time changes in consumer opinions.

How Artificial Intelligence is Changing Human Resources and Hiring

Human resources is an industry as old as the advent of businesses. The focus of human resources is the people element of an organization. The people in a business can cause an organization to succeed or fail, which explains the importance of the human resources industry. The success of the human resource department directly reflects in the performance of a business. The industry should, therefore, keep abreast of developments, which allow them to attract the best people for a business. A workforce can be the competitive edge for an organization.

In many parts of the world, unemployment is rising. The human resources department of most organizations is receiving tons of resumes from applicants whenever they advertise for opportunities. Sifting through the documentation was a nightmare in yesteryears. The advent of AI use in human resources is automating this process. The availability of AI software to sift through resumes sent by applicants is reducing the administrative workload for human resources departments. AI is allowing the human resources industry to focus on the central tasks of interviewing the right applicants for vacant positions. AI resume processing is also allowing human resources to sift through more resumes than ever before.

AI is making tremendous changes within the human resources industry with online platforms budding based on this industry. Platforms like LinkedIn are allowing recruiters to reach potential team players via non-traditional methods. It is not uncommon to find human resource departments sifting through the social media platforms of applicants for a glimpse of who applicants are. The job marketplace is expecting potential workers to have a presence on these platforms. Some vacancies are filling based on the profiles on these platforms. Connections among the workforce in these platforms are leading to opportunities in organizations.

Some organizations are accepting online applications as the primary mode for filling vacancies. The convenience factor is driving this method of getting the best human resources available. Online applications are saving time with organizations that care about environmental issues opting for little to no paper trails in their human resources processes. For organizations with large workforces, the power of online human resource tools is making interactions faster. One can log in to human resources portals and find solutions on a need-to basis. The online systems are making it easier for human resources departments to manage team workers.

Diversity is an element that some organizations have a goal of achieving, which is being aided by AI use in human resources departments. Human resources departments can in real-time, tell if they are within their goal of equality within their workforce. AI is allowing human resources to reach more applicants, which can give companies a global outlook. Human resource departments working in regions demanding strict diversity requirements can keep abreast of these regulations using AI-backed resources. In such environments, adherence may determine the success of a business entity as there are opportunities organizations get based on the diversity of their workforce.

Human resources being a competitive edge in the marketplace has AI to thank, for providing opportunities, for companies to outsource tasks as needed. Smaller companies can get vital tasks sorted without the process of hiring an individual permanently. Many companies are opting to use freelancers for delegatable tasks. AI, in this way, has led to the growth of the freelance industry spanning a variety of jobs. Organizations can, through freelancing, attract highly skilled individuals for required periods whom they would ordinarily be unable to afford. The arrangement offers companies opportunities to leverage the available global workforce to minimize hiring costs.

For the human resources industry, interviews are the tools through which they determine the success of an organization. Depending on the context, the interviewing process can be long-drawn and tiring. Human bias may also affect the effectiveness of the interviewing process. Companies are using AI to negate the human bias component during the interview process. Some companies are choosing to hold interviews over online platforms. The AI-backed interview process is helping human resources sift through suitable candidates with some organizations meeting the top candidates for the final interview. The result is a standardized process that is giving companies a better chance to add the best possible available human resources to their team.

How Artificial Intelligence Is Changing the Sales Process

In yesteryears, the word sales would give a picture of an individual going house-to-house following up on sales prospects. It was a job not for the faint-hearted with rejections a staple of the work description. It required lots of traveling with long work hours expected from salespeople. For most businesses, the sales process is the driver of their existence. Without sales, businesses become extinct. The sales process is, therefore, a fundamental business process that organizations focus on regularly.

Artificial intelligence is changing how organizations are approaching the sales process. Tools backed by AI are now a staple of the business sales process. The Customer Relationship Management (CRM) tool is an AI-backed software that organizations are incorporating to make their sales process more effective. The benefits of these tools include faster responses and better follow up to sales leads. Teams can interact remotely, making better decisions for organizations. Companies are finding it easier to follow up with their sales personnel on the ground using the software. Organizations can access sales data on a real-time basis allowing faster reactions to the market dynamics.

Organizations can now carry out a complete sales process online, for example, through e-commerce stores. Some businesses are purely online while others are taking advantage of the opportunities, and adding the e-commerce component to their brick and mortar stores. The effect is organizations can reach more customers at minimal cost, which translates to higher operational efficiency for businesses. The advent of AI in the online space is allowing for the rise of new sales promotional methods like the use of online bidding. Companies can access the global market using online platforms, therefore, increasing their potential market share.

Artificial intelligence in the sales process is giving rise to the use of applications (abbreviated as apps) in promoting sales. Many companies now have apps that potential clients can download onto their devices through which they can access varied services, including making orders. Such applications can give companies access to information on potential clients that they would have no other way of getting. With the data, companies can personalize the sales process to an individual level, which gives organizations greater accuracy of sales impact. Businesses can use the data as a competitive edge in the marketplace as they create products that will resonate with their potential clients.

For businesses to succeed, they must make a profit enough to sustain the enterprise. Some companies allow for credit, whose basis is the belief that the buyer will be able to afford the payments in the future. Artificial Intelligence is providing tools for businesses to analyze, in advance, whether an individual can pay for their services or products in the future via their use in the credit report systems. The systems have their basis as previous data, which is a component of artificial intelligence. Companies can use these systems to take calculated risks while taking advantage of opportunities.

AI-backed tools are allowing organizations to be accurate in their targeting of sales prospects. For companies selling online, AI devices can give data based on defined algorithms by an organization. The result is minimal sales cost for the maximum impact attainable. The efficiency of the sales process and the coverage is increasing due to these AI-backed tools. The data generated from artificial intelligence is helping companies determine where to focus their sales energies on, for example, in terms of geography, interests, and competition. The data can reveal opportunities organizations previously unknown to companies.

Companies are tweaking the sales process by using AI to lead their decision making. Data collected from artificial intelligence tools is devoid of human bias, which allows businesses to make objective decisions. Accessing the data can be in real-time, which can allow the synergy of the sales process and the dynamics in the marketplace, for example, tweaking of sales promotions as per market response. Companies using AI to make decisions can access new opportunities. They can compare data from other business aspects which can expose trends that can create new avenues and markets, through which, they can sell their products.

Chapter 4: Focus on Strategy and Evaluation

Businesses are opting to incorporate AI into their overall company strategy as a means of staying competitive within the marketplace. Some are going an extra step and creating a separate AI strategy. Businesses are using AI to create a competitive edge over other market players. In highly competitive industries, given the globalization of the market place, AI could be the tool that keeps a company afloat. Some companies are using AI to achieve their overall strategies. Strategies are data-dependent with AI being the link between the two components. History is not lacking examples of enterprises that have collapsed due to a lack of adopting AI into their processes.

Strategy as a process can be broken down into three steps that are planning, implementation, and evaluation. A business should consider these steps to achieve the full potential of AI utilization within a business setting. Plans should be in place evaluating the actual impact of AI use in a business relative to the expected projections. Going through the steps will allow a business to identify gaps and come up with solutions to address the discrepancies between reality and expectations. The steps should be in a constant state of action to make adjustments as quickly as possible.

Strategy First: Plan Your Strategy

Every business worth its salt has a strategy in place. It acts as a roadmap that helps a business focus on its overall vision. The end goal should be a success as pictured in the mind of the business owners. Writing the strategy is an element that helps solidify the map towards the end goal. The process of coming up with a roadmap can help a business deal with issues unthought of when the business vision was just an idea. Businesses should opt for the path that takes them fastest to their end goal while maintaining the integrity of their end goal.

Artificial intelligence is versatile when it comes to strategy as it can be the strategy itself or a tool to achieve the same. Some businesses are using AI as a means to edge out the competition from the marketplace via, for example, utilizing AI to expand their customer base. Others are choosing to center their overall strategy on AI. The latter option portrays a company as adaptable to technological change. Each path has its benefits and challenges with the choice depending on what works best for a business entity. Issues like adaptability by the potential clients of business to AI technology is a consideration when a company is determining which path to follow.

Creating a strategy incorporating AI in whatever form is a process that requires a holistic approach. The expected end impact and its ripple effect on departments making up the business unit is a consideration. Each business unit can involve itself in this process of strategy creation. Having varied views allows for brainstorming that promotes creativity, an element of creating a unique product offering in industries facing high levels of competition. Any weaknesses noted during this period can be a chance for businesses to consider how to turn them into strengths for the business entity.

The creation part of a strategy requires decisions that are best to base on research. Here, artificial intelligence can play a vital role as it lacks the weakness of human bias that can lead to subjective decision-making processes. AI tools are in use by businesses to retrieve data that is assisting in decision making. Looking at data, particularly from different departments, can help in revealing opportunities aligned with the overall goal of the business that can end up being a new source of business for an organization.

When coming up with a strategy, the sustainability question is an element that requires an answer. The business can focus on pursuing strategies that they can afford. The use of data can help in determining what is within sustainable budgets and what is not viable at present. This process allows for the prioritization of strategic roadmaps within an organization. Companies making such decisions on data are less likely to run into accusations of subjectivity among team members. The effect is a synergistic push towards the goals of a business which may hasten the time taken to achieve the intended vision.

The launch of a strategic plan in a business setting should involve all relevant departments to promote a sense of synergy among team members. Business units should prepare for the launch in advance, allowing for changes where needed. When team members feel a part of the process, they are more likely to take ownership of delivering their portion of the roadmap to success. The launch is best done at the same time across all departments so that the impact concerning teething problems is identifiable simultaneously. This method allows for a faster turnaround time of sorting issues and portrays the company to business customers as unified.

As the initial implementation of the strategic plan is going on, a team can be on stand-by to deal with arising issues over time. The stage of strategy implementation is critical as it beats logic for a business to spend time going through the initial steps of strategy without implementing it. Follow up at this stage is helping companies in determining the real effect of the AI tools introduced into the business processes. Resistance to change at this stage may occur as team members may feel the need to go back to traditional methods.

As the strategic tools are in use for longer, the resistance to change will start declining. The benefits of AI to the business processes, particularly how it augments the function of team members is the key to fighting the resistance to change. Once the workforce can see how AI frees them to focus on their core tasks, they are likely to embrace its use. Businesses can address the fears of embracing AI by pointing to the benefits at an individual level. Connecting the benefits to what matters to each team member breaks down the resistance cycle initially expected.

Evaluation as a step in developing strategy is assisting businesses in making adjustments required for transition into AI-backed processes. Continual evaluating is encouraged for organizations not to lose focus on achieving their end goal. Any changes in the market affecting the roadmap is identifiable at this stage of strategy development. Some changes require businesses to change their roadmap and even their end goal. Having a strategy in place is helping organizations in making changes in a dynamic marketplace, helping them avoid becoming obsolete. AI tools can keep information current helping enterprises survive any market onslaught. AI can be a determinant of whether the strategy of a business succeeds.

Evaluate Marketing Tools and Technologies

The AI tools and technologies available in the market are varied, with each offering different benefits. Businesses need to carry out an evaluation process to determine which devices will align well with their overall goals. Having AI that is not in tandem with the roadmap of an organization can lead to wastage of much-needed resources. The effect may be resistance by the workforce in the future as the business attempts to introduce other tools. The evaluation process should be systematic to ensure objectiveness. All stakeholders should play a part to gain their trust, which is critical during the transition process.

Businesses can research the available tools by evaluating the tools currently in use in the marketplace. Reviews from entities already integrating AI into their marketing can give a glimpse of how the tools are working on a day-to-day basis. Going this route allows businesses to choose AI marketing systems already in use. They can opt to incorporate better models of AI marketing tools than those in the market, giving them an edge over the competition. Taking risks on novel AI technologies can pay off if the impact is positive in the long run. Some businesses are becoming market leaders by opting for risk-taking on new AI marketing technologies.

When choosing AI marketing tools and technologies, businesses should opt for those that do not have a bias component or have a minimal bias. The accuracy of the results can be affected by the same, which can result in businesses making decisions that can lead to losses. Marketing tools and technologies having an ability for deep learning are preferable. These require less programming and can give insights that are independent of human bias. Deep learning marketing tools and technologies can learn from previous experience as they can form artificial neural networks.

Marketing tools and technologies having predictive ability allows businesses to react faster to a dynamic marketplace. In a highly competitive global market, such AI software can give companies a competitive edge. Consumers today are demanding and spoilt for choice, making reaction time a critical component of attracting much-needed market share. The predictive factor of the marketing tool in use should be adjustable with those able to predict from the point of drivers of changes being better than those giving predictions from previous inputs. AI marketing tools with machine learning ability are best.

Evaluation requires businesses to monitor the tools that are working well and the extent to which they are fulfilling company expectations. The analysis will allow a company to determine whether an investment in a marketing tool or technology was worthwhile. The report from the evaluation can indicate what areas the business needs to work on and which ones can they bank on to move towards their overall vision. The gaps noted can be filled in by the human workforce creating a synergistic process in the marketing departments. The monitoring should be continual for real-time adjusting.

As with all human inventions, businesses will face challenges as they adapt to AI marketing tools and technologies. Companies should not get discouraged by such scenarios but instead, use them as opportunities to improve on their transition to using AI marketing tools. Choosing to forfeit transitioning to AI marketing tools and technologies can mean the demise of a business entity. Some companies have become obsolete based on not adjusting to the marketplace demands fast enough. The examples are many across many industries with household brands disappearing from the marketplace. Companies should strike a balance between investing in the right marketing tools and technologies and minimizing losses due to wrong investments.

The best marketing tools and technologies for businesses are those aligned with their overall strategic plans. Companies can choose to make their decisions on the marketing devices to incorporate by basing it on whether it will support their strategic objectives. This method allows for objective assessments of available AI marketing tools in the marketplace. The process of choosing becomes clearer when requirements are unbiased. AI can assist in making decisions, particularly when combined with data. Focusing on strategic alignment as consideration for choosing which tool to invest in can remove biases from decision making.

The best marketing tools and technologies lacking the characteristic of ease of use will create resistance from the workforce. A user-friendly interface can be the deciding factor in the adaptability rate in the business. Marketing technologies with unfriendly user interfaces can translate into added costs through the need for training periods that are longer. Staff may require additional time to understand how the tools will work and their roles in the marketing process. Businesses should look at interfaces from the point of the customer as well as that of the workforce. The interface can determine the reaction of the market to the tools a business chooses.

Businesses should choose marketing tools and technologies that are in alignment with the skill set of the workforce. Outsourcing of support can be an additional cost that enterprises can avoid if the skill set is available within the business. Companies can consider training their workforce in skills that will aid in the utilization of marketing tools and technologies. Companies should compare and decide if outsourcing or internal skill training is more plausible, particularly in the long run.

Businesses should avoid being in situations where the marketing tools and technologies are not in use as they wait for support from external sources.

Chapter 5: The Future of Marketing: Predicting Consumer Behavior with Artificial Intelligence

Businesses are looking for ways to predict consumer behavior as a means of maintaining their competitive edge within the marketplace. AI tools classified as machine learning devices can, based on artificial neural networks, achieve this. Given the amount of data that businesses receive, companies need to make sense of the same in a manner that is beneficial to the business operation. The ability to predict consumer purchases may reduce the turnaround time for consumers to get their preferred products. Predictions anticipate client needs in advance, which may be a crucial differentiating factor for a business within the marketplace. Deep learning is a characteristic of AI tools that can predict.

The advantages for marketers in predicting consumer behavior are numerous, including preventing potential loss of sales. Predictions can give a business a sense of the right timing concerning stock replacement. Companies can, therefore, reduce costs associated with inventory, for example, by reducing the time stock spends on the shelf awaiting purchase. Marketers with predictive information can optimally target promotional messages to potential clients influencing their purchasing decisions. For organizations able to combine data plus artificial intelligence, the opportunities for marketers are numerous. In the long run, the benefit of businesses preparing using AI may surpass the investment cost.

Companies able to predict consumer behavior can leverage the same for their sales growth. Predictive data for business use is via varied sources including, social media platforms that potential clients use. The data reveal trends that may be useful to a business entity. When sourcing for data, businesses should ensure alignment with applicable regulations. Clients should not feel their privacy is not guaranteed. Such revelations may be to the detriment of the company as potential clients protest these actions by not purchasing from the affected companies. Consumer behavior data gives companies a glimpse of what is of value to clients.

AI tools can collect this data digitally, which removes the labor-intensive nature of traditional means of data collection. This method is less prone to human bias, which increases the accuracy levels of the data a business receives in this manner. The best AI tools focus on the drivers of consumer behavior as opposed to predictions based on previous inputs. Focusing on the drivers allows companies to be aware of the consumer choices, sometimes even before potential clients are aware that they need the products and services on offer. The consumer behavior data collection can be from personal devices.

Businesses looking to grow their market share should use predictive data by AI tools to align their marketing strategy. Decisions made using data is objective as it misses the weakness of human bias. Predictive information is useful for businesses to adjust their marketing goals in alignment with the dynamic nature of the marketplace. AI data provides information for marketing that would take longer if left to the human workforce. Given the complexity of what drives human behavior, AI can be the source of information businesses have been missing to move to the next level in the marketplace.

Factors to consider when thinking of predicting consumer behavior include levels of prediction, big data and AI, and AI tools for predictive purposes. Each of these elements plays a critical role in predicting consumer behavior, which plays a role in effective marketing by business entities. Businesses should make choices based on their internal needs and how aligned the tools would be to their strategic intent. The consumer behavior of interest includes emotions that some consider as an element that determines human behavior. AI is improving its capability to identify and predict human emotions, which gives the tools insights previously undiscovered.

Levels of Prediction

Prediction levels concerning consumer behavior depends on a variety of factors. These may include the ability of the AI tool, particularly its deep learning functionality level. AI tools that have deep learning capability are more likely to predict consumer behavior to a higher degree of accuracy. The deep learning functionality is dependent on the AI's ability to form artificial neural networks. The data available to a business entity determines the levels to which AI can predict consumer behavior. Businesses that can access more data based on the industry they belong to like customer care companies and retail stores are at an advantage of predicting consumer behavior.

The regulatory environment within which a business is determines what level of data access is legal. Highly regulated environments may limit to a great extent the kind of data companies are allowed to access from potential clients. The levels of prediction achieved of consumer behavior should tally with the strategic objectives of the company investing in AI tools. Unnecessary expenditures should be limited to functionalities relevant to a company's level of need for data. Businesses should focus on AI tools of prediction that are within their investment budget so that the company does not undergo unnecessary monetary stress. The long term effect of the purchase on the company's financials should be the standard for making the decision.

The skill levels at the disposal of the company is a factor that will determine the levels of prediction the business can achieve in the long term. Training of in-house staff and outsourcing of relevant support tasks should be a consideration by businesses incorporating AI tools for prediction purposes. Human behavior is considered complex, and predicting the same is sometimes impossible, even with the best AI tools. Companies choosing to utilize AI tools to predict human behavior in their marketing exploits should be aware of the limitations of the technology.

The machine learning capabilities of AI tools determine the levels of predictions available to businesses choosing to use AI in predicting human behavior. AI tools that can collaborate with supportive technologies are better placed to predict consumer behavior. Other tools that can support AI include those used for analytical purposes and the human component, the latter of which has strengths that AI is still lacking. The levels of prediction of consumer behavior within the marketing context include intent analysis, category patterns, and future actions.

Intent analysis, also known as emotional analysis, refers to the prediction of the emotions that underlie consumer behaviors or the intent behind their actions. Marketing that connects to the emotional state of consumers is likely to impact clients more than one that deals with non-emotional cues. The emotional connection allows business entities to garner a competitive edge within the marketplace. When companies connect emotionally with potential clients, they get their loyalty, which can assist a business increase its market share. Knowing the why behind a consumer's decision can help companies connect better with potential clients in the marketplace within which a company finds itself.

Category patterns are all about segmentation, with this level of prediction allowing businesses to group potential customers that react similarly. The ability of an AI tool to segment populations will allow for the creation of marketing promotions that have a personal feel to a segment of customers. Segment ads can give a feeling of belonging to potential clients, which may earn their loyalty in the long run. Segmentation allows for a better understanding of the categories identified by AI marketing tools as the segments undergo more study. Companies can avoid mistakes that may, for example, cause offense which may cost a business potential clients.

AI tools that can predict future actions of consumers are valuable to businesses in the long run as the entities can prepare for client expectations in advance. This prediction level can be the key for organizations to influence the future behavior of consumers as they focus on the drivers of their actions. Predictive abilities by business through the use of AI tools can propel a company to an influencer position in the marketplace. The lessons this level of prediction can give to businesses can help in aligning strategies to consumer expectations.

The AI industry is focusing on creating tools with predicting levels of future actions, and the expectation of businesses will be more accurate reports. The goal of AI concerning future predictions is doing so in real-time so that organizations can keep up with changes in the marketplace as they occur. The effect will be a change in how marketing will look like in the future with businesses changing their sales techniques. Marketing tactics will also shift as more information on consumer behavior comes in a timely fashion to businesses. AI tools will be a critical component for marketers looking to connect organizations to potential clients within the marketplace.

Big Data and Artificial Intelligence

The backbone of artificial intelligence is big data, which gives AI the foundation on which, it develops artificial neural networks, that determine the levels to which, AI can predict the future. AI outperforms humans in processing large volumes of data that is an element of future predictions. AI tools can process information at a faster rate than humans and analyze the same with more precision than a human workforce. The processing of big data can lead to the revelation of patterns that can give insights to businesses on market dynamics that they can turn to opportunities.

The precision levels of predicting consumer behavior rise proportionally with the volumes of data available for analysis from varied inputs. Businesses can understand consumers better and can predict their future actions with more clarity than when the sample data is minimal. Combining the elements of large volumes of data and artificial intelligence tools with deep learning capabilities is the future of predicting consumer behavior. With the right amount of data combined with the right AI tool, businesses will be able to make predictions up to months in advance. Organizations can use the results to prepare for the needs of potential customers by, for example, purchasing adequate stock for their anticipated requirements.

Such data can help businesses minimize loss due to the stocking of obsolete products that will no longer meet the needs of the market in the days ahead. Big data can realistically undergo processing via traditional methods due to the complexity and volume of the expected results. The data is dynamic as it is under continuous update as various sources act as inputs to the systems tracking consumer behavior even in real-time. Businesses are adopting big data and AI to keep ahead of the competition with organizations that delay risking becoming irrelevant in the marketplace.

Big data and AI can give businesses forecasts of up to a year from the present time, which gives companies requisite time to adjust to market changes. The predictions focus on the underlying drivers of decisions like human emotions and the causes of the intent behind consumer decisions. The sources of big data for a single business can run into the thousands, which give results that can be highly accurate. Organizations with access to such information can tailor their products almost to a real-time basis, sometimes turning into influencers of consumer decisions in the process.

Businesses can use the data and results from AI to become a part of the culture of consumers in the marketplace giving them a competitive edge over other market players. Big data is negating the function of focus groups as AI is allowing for the processing of large volumes of data. The unbiased nature of data fed into AI systems is allowing companies to make objective decisions faster with more assurance. AI outperforms the human workforce in analyzing big data as it can work with varied data types, whether structured or unstructured.

AI tools have algorithms by which they analyze big data into insights that businesses are using to take action in the marketplace. Source of big data for use by AI tools for turning into insights for business entities include social media platforms and trending news topics. Big data and AI is in use by organizations to adjust their strategies in the marketplace as pieces of information come in from varied sources. Businesses that can anticipate how consumers will react to situations can tailor their brand to be part of the positive side of culture. AI via big data can act as a research laboratory for businesses looking to introduce new products and services into the marketplace.

For AI with future action prediction ability, the opportunity to get a feel of how consumers will react can be priceless for a business as it enters new markets. The kinds of AI that can combine with big data include pure AI and pragmatic AI with the latter combine various technologies to give insights. The choice between the AI types is dependent on a business and its vision for the results of predicting consumer behavior. Organizations can enhance the experience of their consumers when they utilize AI to predict the future actions of their clients.

Big data and AI can give results that allow businesses to come up with products and services that potential customers can relate to easily. Relevance breeds loyalty in the marketplace, an ingredient that can propel once unknown brands to levels of influence above the competition. Segmentation through the identification of patterns from big data by AI allows businesses to present marketing promotions that are applicable at individual levels. Big data and AI can help companies provide support to their customers in real-time. The result is a positive experience for the customer who may share their view of the company, therefore possibly increasing the market influence of the brand.

Artificial Intelligence Tools for Prediction

Artificial intelligence tools come in many forms with the variety based on their functionality, particularly in terms of deep learning and machine learning abilities. Businesses should choose the AI tool in the context of their strategic intent as AI can help a business reach its overall vision faster. The best kind of AI tool is one that can integrate with other systems already in use within the organization. This feature helps in the transition period reducing the possible issues that can arise during the implementation stage of AI tools.

API.PI, Google Cloud, Infosys Nia, Microsoft Azure, Premonition, TensorFlow, and Wipro HOLMES. API.PI is an AI tool that allows businesses to customize the software to their brand while using natural language in its functionality. It combines its functionality with data from varied sources including encyclopedias, information on flight schedules and weather patterns. Businesses whose marketplace is driven by such data can utilize the AI tool to predict consumer behavior. The insights given will allow the organizations to be better prepare for the expectations of their clients.

The AI Google cloud is a prediction tool that bases its functionality on machine learning capabilities allowing it to identify patterns in the data of a business. It can then predict the future, for example, categories based on real-time data giving suggestions to organizations on actions to take. Its machine learning capabilities can assist companies in detecting mails the entity can categorize as spam. The tool can work with varied computer languages, including Java and Python, which make it versatile. Businesses can, therefore, integrate it with their in-house tools, making it a preferable choice for organizations looking for a smooth transition into AI use.

Infosys Nia is an AI tool having capabilities to predict described as knowledge-based, which in combination with its machine learning capabilities can allow businesses to automate processes. Businesses using Infosys Nia can employ a culture of innovation, riding on the information presented by the tool, as it analyzes company data. Infosys Nia can solidify fragmented processes to give companies insights that may reveal new opportunities. Businesses can use the machine learning abilities of the tool to work on their systems continually which companies can use to stay competitive in the marketplace.

Azure by Microsoft, is an AI tool, with abilities to predict, that employs machine learning analysis, and is cloud-based, making it accessible, regardless of a company's geographical location. Azure presents analytical results in a simple way, which allows businesses to make decisions faster. Azure is a web service that can be mobile-enabled, with varied applications including business intelligence, digital marketing, and e-commerce. Microsoft Azure is suitable for businesses looking to gather information from different sources continually to stay relevant in the marketplace. The predictive AI from Microsoft can work with different operating systems making it a viable option for businesses looking for AI tools that are versatile.

Premonition is an AI tool that is specific to an industry that is the litigation industry where it gives information on attorneys and the cases they have handled. It uses algorithms, that are of interest to their clients, like how many cases an attorney has won, what rate an attorney charges, and what kind of cases they have presented. It is a database of litigations that are analyzed, continually giving clients real-time information that they can use to determine which attorney to work with on their cases. The information allows clients to question potential attorneys on whether they are the right fit.

TensorFlow is a predictive AI tool that helps businesses predict consumer behavior, which is a critical element of staying relevant in the marketplace. TensorFlow is an open-source program that focuses on creating machine learning models that businesses can use for prediction purposes. TensorFlow works through the numerical computation of data flow in the form of graphs with nodes and edges of the graphs having different meanings. Businesses can use TensorFlow to build neural networks based on data from different origins that organizations are using to analyze consumer behavior. The AI can translate languages using the neural networks formed through machine learning capabilities.

Wipro HOLMES is a predictive AI with a myriad of capabilities, including deep learning, processing of natural language, and semantic ontologies. These abilities help this particular AI possess cognitive characteristics that are critical in the development of robots, drones, and visual computing. Companies can use Wipro HOLMES to create virtual digital assistants that have cognitive abilities. The AI can automate processes, including the ones involving the cognitive characteristics. The result is businesses can continually improve on the quality of their artificial neural networks. The algorithms used in Wipro HOLMES include deep learning and genetic learning as components in developing its cognitive characteristics.

Chapter 6: Chatbots and Autoresponders

Artificial intelligence is changing the way businesses carry out tasks that they consider repetitive and mundane, with organizations looking for ways to automate these processes. The AI tools organizations are opting for are varied and depend on factors that may align with the size of the business. The benefits of the AI tools are many, with some needing little AI knowledge to incorporate into business processes.

One of the reasons why businesses opt for these AI tools is the ability for the devices to automate processes that take the human workforce from focusing on the core mandate of the organization.

There are various AI tools that businesses can use, to computerize their recurring tasks, including chatbots, and autoresponders. The organizations make a choice depending on, which AI tool best suits their companies. Some companies choose AI tools as a way to reduce their reliance on human labor, for example, if the business relies on seasonal orders.

The AI tools used for automation do share similarities with some not being charged by the companies that develop them for use by those interested in utilizing them. The commonalities can translate to an identical range of advantages depending on how businesses choose to implement them.

The benefits for the AI tools chatbots and autoresponders may include the ability for businesses to increase the scale at which they can respond to potential customers. The organizations increase their efficiency levels by using the chatbots and autoresponders. Companies can integrate some chatbots and autoresponders into their social media platforms through which they interact with potential clients.

The processes that the AI tools chatbots and autoresponders can automate are varied, including interaction with customers on a real-time basis. Some are suitable for scenarios whereby the human workforce of a business is unable to respond immediately to customer inquiries.

Businesses are utilizing the AI tools in augmenting their marketing efforts within the marketplace as they form stronger bonds with their potential clients. Some companies are integrating the AI tools giving potential clients a better customer experience that is synergistic. The implementation of these AI tools by businesses can be a competitive edge in a marketplace that is digitally dynamic.

Chatbots

The abilities of chatbots vary depending on the creators with characteristics depending on the implementation within a business. They may have auditory and textual aspects with the specifics depending on the range that an organization may require for their processes. The focus of chatbots is to replicate humans in the actions they are in use in the marketplace. The Chatbot software can utilize natural language in performing their functions within business processes. Chatbots can interact and integrate with other applications that organizations are using within the marketplace.

Businesses are personalizing chatbots in their interactions with potential clients through this AI software platform to build their brands. Organizations are working with Chatbot creators that are allowing companies to create an identity that distinguishes them from the competition. Businesses should be aware that there are chatbots that do not have AI capabilities, which limits their range of actions. Chatbots perform actions by relying on algorithms that are preset depending on the specific requirements of a company. The requirements include the level of interaction that a business requires, from the Chatbot, for example, before a human interface takes over the conversation.

The advantages of chatbots that businesses can take advantage of include their ability to work with autoresponders, which are part of the AI software devices. Some organizations have increased their sales by using chatbots in their interactions with potential clients looking for sales information. Chatbots are versatile in their use in business processes, with their ability to function across departments, for example, sales and marketing. Chatbots are convenient to use with businesses varying their functioning depending on the end goal. Some companies are using chatbots to deal with FAQs (Frequently Asked Questions) interactively capturing the attention of their clients.

As with all technology, there are challenges that businesses are facing with chatbots, including the software giving irrelevant answers to potential clients. Some companies are reporting chatbots that are giving incomplete answers to their customers, which may lead to loss of business. The question of monetization also comes into play, with some companies struggling to find ways of creating sales from the chatbots. Some chatbots tend to annoy clients with their programming giving a feel of talking to a machine vis a vis a human. Chatbots, although classified as AI, is not yet considered as passing the Turing test, which is a standard of determining whether a machine can think on its own.

What are Chatbots?

Chatbots are a type of artificial intelligence software that can be in use either in an auditory or textual context that automates the process of interaction. Depending on the Chatbot AI level, they can mimic to a great extent the actions of a human that one is interacting with, in a conversation. The AI class within, which chatbots fall under, is conversational AI because they are in use, in the context of communication. The algorithms used in chatbots include pre-programmed user phrases either in the form of texts or sound.

Chatbots come in many forms with their classification depending on their abilities. The main categorizations are AI and non-AI with the latter denoting those that do not have machine learning capabilities. The non-AI category of chatbots is sometimes referred to as scripted. Another categorization model classifies chatbots based on whether they work alone or in combination with other software like messaging applications. There are chatbots known as menu-based which have a flowchart kind of design that leads one to various points depending on the questions asked. Some chatbots work across a variety of platforms, therefore, classified as multi-platform.

The uses of chatbots vary with the most common application being the scaling of customized customer care within the setting of a business interacting with clients. Chatbots are also in use for gathering the information that organizations are looking for to enrich customer experiences. Some chatbots are in creation for entertainment purposes with users interacting with the AI software for recreational purposes. There are chatbots whose use is to assist individuals in performing tasks that they assign to them. Chatbots are in use to substitute how communication is occurring between businesses and clients with this AI software being more interactive than some traditional methods.

Chatbots are in use in a variety of industries, including e-commerce, where it is allowing potential customers to interact with the products and services on offer in the business. The marketing industry is using chatbots for a variety of promotional campaigns across different platforms. The sales industry is on board with using chatbots with the interaction with potential clients allowing for customization. The travel industry is using chatbots in booking for hotels and airline trips for potential customers and directing clients to partner stores. The healthcare industry is using chatbots to gather information from clients before connecting them to a human interface.

Tools and Platforms for Chatbots

The tools and platforms for chatbots come in many forms with varied benefits depending on their capabilities, giving businesses a choice. Organizations should choose the tools and platforms for their chatbots that most closely mirrors their path to achieving their strategic objectives. The tools and platforms are in use for building chatbots with some being easy to use, and others are more complex. Businesses are using the tools and platforms to create chatbots that are helping them maintain their competitive edge. They are also in use in creating chatbots for personal use.

There are tools and platforms for chatbots that require the creator to have coding skills, while others do not require the same. The platforms can be open source or can require payment for use by private entities. Some tools and platforms for chatbots include Chatfuel, Botsify, and Flow XO with each having varied benefits. Chatfuel can integrate with Facebook messenger and does not require the skill of coding with the platform offering an option for use at no charge. Paid versions in the platform give more information to a business with functionalities like data management.

The Flow XO platform for creating chatbots also boasts of one not requiring coding skills to use their tools and can be used in a variety of platforms. The ability for this tool to work across many platforms makes it versatile, which can give businesses convenience if they want to utilize across varied platforms. The platform allows for integration with other applications and is allowing incorporation into websites. The features allow for businesses to let their customers share their Chatbot with others, which can create a referral system for organizations. The platforms allow for one to create several chatbots, which creates more avenues for use by their potential clients.

Botsify is a tool and platform for chatbots that works across Facebook and on websites making it a multi-platform tool and, therefore, versatile. It can integrate with other applications that a business can use to communicate with their potential clients making its incorporation into company processes easier. This platform allows organizations to merge the technology of Chatbot with a human interface. The integration may prove to be a competitive edge for a business in the marketplace as some clients may prefer talking to a fellow human. There is a free version available, and the platform can allow companies to create conversational forms.

Promote Your Chatbot

Businesses can promote their chatbots in many ways depending on the strategic objectives the organization is looking to achieve with the AI technology. Companies should prepare for the promotion of their chatbots in advance, taking into account the benefits that are important to their potential end-users. How the target consumer of a company will perceive their interaction with a Chatbot will determine the effectiveness of the AI software. The preparations should be extensive with businesses being ready to spend the required time to present a product worthy of their brand.

Businesses can promote their chatbots by adding buttons on other platforms that they have a presence on, for example, on their websites. The buttons leading to their Chatbot experience should be in a visible place that potential clients are likely to see to increase users. The chatbots for businesses should contain exclusive codes either as a messenger, QR Code, or a URL Code. The QR codes best function when organizations place them in non-clickable areas like in billboards and posters which allow potential clients to scan the codes. Businesses should note that the messenger code is specific for the Facebook platform.

Businesses can promote their chatbots through the use of plugins that can connect to the presence of the organization in other platforms, for example, on websites. The plugins will direct potential customers to the chatbots on the platforms where the Chatbot lies. One can take advantage of the opportunities presented by bot stores that work with a company to promote their Chatbot with businesses choosing the store that aligns with their strategic interests. Companies can create a story around the Chatbot to keep customers interested in using the AI-backed platform. Some businesses are creating landing pages for their chatbots as a way of explaining to potential customers how the Chatbot can assist them.

The landing page can be SEO (Search Engine Optimization) optimized to increase the reach of the business, encouraging more people to try out their Chatbot. Companies can use their social media pages to promote their chatbots, giving detailed explanations showing how easy it is to use the same. Potential customers may choose to give it a try when they see visuals on how to make use of it. The businesses can use the option of paid promotions to take advantage of the opportunities to create customized targeted messages.

Reduce Customer Service Workloads

The benefits of chatbots vary with one of the main advantages being to reduce customer service workloads as it automates business processes. In decreasing customer service workloads, businesses can reduce the costs related to interacting and following up with customers. Some sources say that chatbots can reduce customer service workloads by up to one third depending on how a business implements the AI software. Chatbots reduce customer service workloads by streamlining the processes in the department. Companies can focus on delegating the customer service tasks that are administrative as they are usually repetitive and mundane.

Reducing the customer service workload allows the human workforce of a business to focus on tasks that are too complex for the chatbots a company decides to use to automate their processes. The human workforce can then focus on creating engaging interactions with potential clients. Chatbots allow for scaling in terms of the amount of customer service workload a business can handle as they can serve multiple customers simultaneously. Chatbots reduce the customer service workload by allowing for the establishment of customer service for hours beyond the working hours. Businesses can achieve faster turnaround times by reducing customer service workloads.

The techniques that businesses can employ in reducing customer service workloads include creating a Chatbot interface that gives clients links to FAQs (Frequently Asked Questions). These help in filtering repetitive inquiries that the business has solutions to that chatbots can handle. For more engaging chatbots, entities can choose to use avatars and emoji that can create an emotional connection with potential clients. The chatbots should not overwhelm the customers as they may opt to walk away from the business or interact with the human workforce. The result is the chatbots will not have a maximum effect in reducing customer service workloads.

Businesses should consider using chatbots in reducing customer service workload as an opportunity to enrich the customer experience with their brand. Companies are revolutionizing their customer service workload processes using this AI software that can multitask beyond the human levels. Chatbot integration with the customer care team is now possible in a seamless manner allowing for better performance. These tools are allowing businesses to be available in sorting out their customer needs all through the day and night. In this case, they act as a continuation of the customer care workforce after working hours.

Autoresponders

Autoresponders, just like chatbots, are in use by businesses to manage their customer service processes that they consider as routine and mundane. The autoresponders allow companies to free their human resources within the customer service department to focus on more complex tasks. Such tasks left for human resources may include objectives requiring the cognitive element of individuals that some machines lack. Companies are choosing autoresponders that can integrate with other applications that may already be in use in their business processes. They are also looking for autoresponders that allow for seamless human-interface incorporation. The autoresponders are allowing companies to scale their capabilities within their customer service departments.

Autoresponders are achieving their tasks by automating, which is allowing for multitasking and scaling of customer service processes by businesses. Businesses are personalizing the automation process to reflect their chosen pathways towards their strategic objectives. The complexity of the automation processes can differ among organizations depending on the capabilities of the Autoresponder chosen. Autoresponders with machine learning capabilities can achieve higher levels of automation as they have a sense of cognitive abilities. Businesses need to determine the optimal level of automation they require taking into account the intricacies of their industries.

Autoresponders can be in use by businesses to achieve segmentation or targeting of potential clients as they can be in use to focus on particular customer segments. Some organizations choose customer segments by allowing for signing up for autoresponders. Businesses should consider customer privacy policies when integrating autoresponders to avoid breaking regulatory requirements or create a negative experience for potential customers. Potential customers want to feel they are in charge of the communication process with businesses they choose to communicate with on various issues. Companies can achieve segmentation by giving choices to potential customers on the type of autoresponders they prefer.

Autoresponders are useful for processes that do involve FAQs (Frequently Asked Questions) as some businesses are incorporating them into the AI software. These questions are generally repetitive and asked in similar forms by potential clients seeking answers for their inquiries from companies. Businesses can tailor FAQs as they communicate with their clients, making changes as required. FAQs assist businesses in reducing their customer service workload as clients can take advantage of the self-service aspect of the responses. Some autoresponders present FAQs in a flow chart system leading customers to the appropriate response in an interactive manner as opposed to the traditional search methods.

What is an Autoresponder?

Autoresponders are AI tools limited to the use of emails as the point of interaction with potential customers who reach out to the business. The response in the form of an email is in a standard format that is pre-programmed and usually does not require an answer. The complexity of autoresponders varies depending on the requirements of a business regarding interacting with potential clients. The initial point of contact with an Autoresponder can include when a potential client subscribes to receiving information from a company. Within the subscription process, businesses can share the cycle of email deliveries to expect with the potential client.

Various types of autoresponders are available in the market with companies opting for those that align with their overall strategies. Companies utilize strategic objectives as triggers setting off Autoresponder systems incorporated into their processes. Sequence emails are a type of Autoresponder where the sharing of particular topics or related topics is through sending emails at regular intervals. These work well in keeping subscribers engaged and can be used to lead them to affiliate websites. For businesses looking to convert interested customers, lead-generation autoresponders may be suitable. These give more information on the products of interest over a period to lead the client to purchase.

The uses of autoresponders are as varied as the goals a business may have, including educating subscribers on topics of interest over a period. Autoresponders are in use by companies looking to keep in contact with their subscribers by sharing engaging content at regular intervals. Some businesses use autoresponders as a pathway to promote their affiliate businesses or products to subscribers. Companies can use autoresponders to share the content of interest to subscribers using a sequence of preprogrammed emails. Companies can use autoresponders to gather input from subscribers, for example, by including survey links within the emails.

Platforms providing autoresponders for personal and business purposes vary in both price and capabilities. The best autoresponders are those that mail delivery platforms do not classify as spam. Autoresponder platforms can be outsourced or set within the server of a company with the latter requiring technical skills. The word Autoresponder is interchangeable with email marketing as the AI software use is generally limited to interacting through emails. Apart from the timing factor, businesses incorporating autoresponders decide on how many emails to send per interaction with the customer.

How to Choose an Autoresponder

There are many options for businesses looking to incorporate autoresponders in their processes with the choices, including outsourcing and in house platforms. The Autoresponder chosen can add or take away from the journey of a business towards its strategic objectives. Companies should consider how the autoresponders align with their goals and their versatility in application. The market provides free and paid versions of autoresponders allowing businesses of different sizes to benefit from AI technology. Organizations should consider their sales cycle when deciding on the

Autoresponder to utilize as the length should direct the frequency of sending the emails.

An organization can consider many factors when choosing the Autoresponder to utilize, including the ease with which they can create an email. The ease of sending the emails is critical with those allowing for presetting of time of email release preferable as they are convenient to use. Autoresponders with tracking capabilities are useful for businesses in determining the effectiveness of the content they are sharing. The tracking results can be applicable in benchmarking how well customers are responding to sent emails. For companies using the emails to promote products and services, choosing autoresponders that provide inline promotional email templates can be preferable.

Another consideration is the number of emails the Autoresponder platform can allow a business to send over a period in the context of the sales cycle of the company. The autoresponders of choice should augment the goals of a company and not take away from it. Its ability to integrate with other applications in use by companies is critical in guaranteeing a smooth incorporation process. The level of segmentation is another consideration for companies, particularly for organizations looking for highly targeted content sharing opportunities. When it comes to automation, businesses should choose autoresponders that limit the need for human action in terms of responding to subscriber options.

Companies can look through reviews by other users to get a feel of what the perception is towards the service of autoresponders available in the marketplace. This action may save a business from investing in an Autoresponder that will disappoint, and that may erase gains in the market. Companies should choose autoresponders that have features important to them and not a myriad of features yet do not apply to their context. In most cases, how long a company that is providing Autoresponder platforms have been in the market is an indicator of their competence. Their interactions with companies handling emails will determine if mails get delegated as spam.

Enabling Autoresponders

Enabling of autoresponders is dependent on the platform a business chooses to utilize in integrated the AI technology into their processes. Autoresponder platforms offer various category capabilities to choose from, including intervals between emails to be sent and details on the text to be included. The periods between the emails are varied, with some platforms giving choices in hours and others down to minutes. In regards to timing, and depending on customer location, businesses can choose the delivery to align with the local times of the clients.

Businesses will need to specify the triggers for the autoresponders taking into account the context of the focus of their strategic objectives. The strategic goals should be the first step in determining to enable autoresponders as the effectiveness will depend on how aligned the expectations are to the capabilities of the AI technology. The objectives should be listed to ensure every possible aspect is covered. All departments within a company setting should be part of the process. Businesses match Autoresponder capabilities, and strategic objectives, at this stage.

Activation of Autoresponder features that match the strategic objectives is the next step with details like the timing of the intervals triggered. Businesses can define the cycle lengths between communication times and update the contact list of subscribers. Segmenting of customer lists is possible depending on the capabilities of the Autoresponder platform the business is using to communicate. Filling of the details of the body of the email message, including using templates as provided by the Autoresponder platform, is possible. When defining cycle lengths, businesses can choose the particular days they want the communication forwarded. Tracking features are enabled once all the details are activated.

Most Autoresponder platforms give detailed explanations on how to set up the AI software in a business context with some providing FAQs (Frequently Asked Questions). Businesses can utilize this portion to maximize the capabilities of the Autoresponder platform they decide to use. One can send a test email to have a feel of what the end customer will receive, which is a good practice to allow for any changes in advance. As customers interact with the Autoresponder platform of choice, businesses can be open to making changes in response to feedback from clients. The main focus is to be responsive to customers in a convenient manner to the company.

Autoresponder Email Message

Autoresponder email messages are communications sent automatically in response to an action by another party communicating with, for example, a business. The content of the Autoresponder email messages are pre-programmed and linked to specific triggers. Pre-programming allows businesses to homogenize their communication with potential clients with organizations using similar templates. The templates can contain different elements with various platforms offering free and paid versions. Autoresponder email messages may be a medium through which a potential customer interacts with a business for the first time. Autoresponder email messages are also known as out of office messages as they are generally in use when one is unavailable in a professional context.

Autoresponder email messages come with a variety of advantages, including reducing customer workload regarding tasks that are repetitive and mundane.

Businesses using Autoresponder email messages can serve customers at a larger scale as compared to working with a human workforce. The result can be an increase in market share as the business can do more with fewer resources, freeing the human workforce to focus on core tasks. Autoresponder email messages also help companies portray a professional image to those interacting with them as it is considered a standard business practice.

Autoresponder email messages come with challenges, including potential clients getting frustrated, for example, in contexts where they need an urgent response. Depending on how the Autoresponder email message is formatted, customers may find the response cold. Drafting an Autoresponder email message that fits across all contexts may be a challenge depending on the industry dynamics. The benefits though outdo the negatives, including Autoresponder email messages acting as an automatic while you were away system of response. Autoresponder email messages can help in dealing with the feeling of frustration when one gets no response.

When instituting Autoresponder email messages, one should remember to provide options on the way forward like alternative contacts. The Autoresponder email message body should be professional, taking into consideration the varied contexts it is meant to serve. Details like how long one is likely to be away can be part of the message body with the actual return date also inserted. Businesses can consider Autoresponder email message branding in maintaining a brand presence across their communication channels. Autoresponder email messages should be brief, with one placing reminders to activate them before being absent. The main goal of the Autoresponder email messages is to inform the communication partner of one's absence.

Automation Rules

The automation rules regarding Autoresponder email messages are specific to trigger matched responses to actions taken by the recipient. The parameters of the rules specified must be spelled out to direct the platforms on the email message to send from the pre-programmed bundle. The rules are specific to business processes with organizations taking into consideration their context. The specifics of the rules are dependent on the platform that businesses use to implement their Autoresponder email messages. The parameters help in handling multiple lists of contacts who should receive different messages depending on their actions.

Automation rules assist businesses in making the process of customer segmentation less manual, allowing them to handle clients at a larger scale. The segmentation should allow for shuffling of the pre-programmed emails sent when a business changes the triggers. The result is organizations can customize their Autoresponder email message body in alignment with changes in the marketplace. Companies should consider the email message body in the context of their overall strategic objectives taking advantage of the opportunity presented by subscribers. Businesses are using the Autoresponder email automation rules to include marketing messages that promote products and services to subscribers.

The automation rules change depending on the industry with some specifying actions based on purchasing decisions and others on information requests. An automation rule can specify an Autoresponder action when someone pays for an item where the message body can contain details of the transaction. For the latter case, businesses can use the rules to send information to clients at regular intervals. Companies should include options to unsubscribe within the body of the Autoresponder email message sent to clients. Some businesses opt to, for example, let a subscriber remove themselves from an email list while others want the subscriber to specify the particular emails they want stopping.

The effect of automation rules depends on the complexity of business operations with the impact proportional to the number of email lists requiring managing. Some platforms classify the automation rules into when, if and then with the first parameter stating when the rules apply. The second parameter will detail the criteria that need fulfilling for the sending of the Autoresponder email message. The platform should allow for capturing of data of triggers and actions occurring based on the automation rules set by the business.

Chapter 7: Scaling Up Your Business: Artificial Intelligence Marketing Platforms

Many artificial intelligence marketing platforms can help in scaling up businesses, including autoresponders, chatbots and business analytics. Each of the platforms come with various benefits which enable companies to move towards their strategic objectives at a faster rate. Benefits include saving time with AI supporting automation, which allows companies to achieve a quick turnaround time. The result is some companies can reduce costs previously associated with manual marketing efforts as scaling is possible with AI. Businesses employing AI in their marketing efforts can manage more activities that are personalized and specific to customer interests. The platforms are evening the marketplace with startups using AI to compete with established businesses.

Emails as a marketing platform that is AI-based is automated and allows for lead generation when businesses use the message body to, for example, share product availability. Having an email as an Autoresponder is improving the customer experience businesses are offering. Automated emails can serve more customers, as it diminishes the laboring requirements of the human workforce, writing an email per customer. Automated emails can contain links to Frequently Asked Questions (FAQs) to reduce the number of customer inquiries. The email message body can contain promotional messages or links to affiliated sites.

Chatbots are in use widely by companies as an AI marketing platform to scale the reach of business in various operations, including promoting deals. These improve customer service as the response is instant, and it covers the questions frequently asked by potential clients. The quick turnaround time exhibited by chatbots can be used by companies to gain a competitive edge in the marketplace. They are convenient to use as most have a flowchart interface that answers inquiries as an interactive session with potential customers. Chatbots are efficient as they are versatile with some working across various platforms available to businesses.

Other autoresponders include text messages connecting to actions taken by consumers as they interact with businesses, for example, after transactions. Texts can contribute to lead generation, with companies including links within the body of the message. Businesses using text messages as AI marketing platforms are serving more customers with less labor required of the human workforce. Companies are using texts to receive feedback from customers in real-time with organizations requesting ratings of services offered or products purchased. The businesses can achieve this through the use of two-way text messaging AI platforms. Companies are including links to promotional activities in messages to customers as they look for opportunities to upsell.

Businesses are using analytics powered by AI to gain a competitive edge in the marketplace with the tools analyzing data from customers. The analysis reports are in real-time for quicker turnaround time by businesses as they seek to align themselves to a dynamic market. These reports are allowing companies to make decisions that are of higher impact as they react faster to changes in the market place. The predictive nature of AI marketing tools that can analyze is saving costs for businesses as they can reduce inventory costs. Insights by AI-powered business analytical tools are actionable.

Businesses can use AI-powered marketing tools that can optimize content to allow for a higher impact in the marketplace at a lower cost. These tools are allowing companies to reach more individuals as scaling is possible, depending on the requirements by specific businesses. The content generated is being distributed by various platforms, including autoresponders, like emails that may require a subscription from interested customers. The AI platforms also act as a source of content for businesses to take advantage of while creating content. Companies can outsource managing content to AI-powered platforms that are helping organizations align messages shared with their strategic objectives.

Social media channels are becoming the go-to AI-powered platform for businesses looking for an interactive experience for their customers. Here, companies are building brand loyalty as customers interact with them on a larger scale with a quick turnaround. Businesses on social media can deal with customer inquiries and concerns on a real-time basis, which may increase their market influence. Some social media platforms are providing organizations with opportunities to sell their products and services. There are AI marketing tools that are facilitating businesses in marketing across various social media platforms at the same time.

Beacons are AI-powered tools that have the unique capability of interacting with customers who are in close proximity to, for example, a brick and mortar store. Businesses stocking items that customers are buying based on impulse are taking advantage of the opportunity that these platforms are presenting. Some are connecting with applications that are sending pop up notifications to clients who are passing nearby stores looking for customers. Companies are sending specific offers based on various factors like time of the day to, for example, invite passing customers for lunch offers. The beacons are also providing data for businesses that are aiding in decision making that is objective and is useful for future marketing activities.

Businesses are using digital ads as an AI marketing platform to reach customers in a dynamic way that aligns with the interests of the target clients. The ads have the capability of personalization as, for example, a potential customer is browsing through an affiliate website. Some AI marketing platforms are providing businesses with an opportunity to sponsor the ads to allow for the segmentation of customer groups. The result is a higher impact of the ads as compared to traditional ads achieved at a lower cost and a larger scale. Digital ads can work across platforms including mobile phones, laptops, and different websites, changing depending on pre-programmed triggers.

Many businesses now have apps that customers are using to interact with organizations in various capacities, including accessing promotional messages. The messages may come on screens as pop up notifications which, when clicked, give more information on marketing activities. The advantage with apps is the amount of data that a business can gather from a customer who chooses to install the AI marketing platform. The companies can use the data to personalize marketing campaigns while gaining knowledge on the network of individuals connected to their primary customers.

Chapter 8: Artificial Intelligence Email Marketing

Artificial Intelligence (AI) email marketing is a part of digital marketing. It has the benefits of relying on machine learning, and deep learning depending on the complexity of the platform an organization is using to share the emails. The goal is sending emails to a potential customer that is personalized, allowing for the interaction to be human-like. AI email marketing is allowing businesses to scale their customer service operations at a reduced cost. Companies can use AI to limit the possibility of the categorization of their emails as spam.

The data from machine and deep learning is in use to align customer interests with business goals. AI email marketing can be used to approach potential customers directly, avoiding the cost of the middleman in marketing. The result is a better use of resources, which can mean an increase in the number of customers a business can serve at any one time.

The automatic nature of AI comes with the convenience not associated with manual processing. AI email marketing is known to portray a business professionally with the marketplace expecting the responses as standard practice. Features of AI email marketing may include efficient send time optimization and effective automation workflow. Each of these facets can assist a business in improving different aspects of their operations. Companies now have various platforms they can choose to institute AI email marketing at their convenience and with varied complexity.

The Convenience of Automation

The convenience of automation is one of the advantages of artificial intelligence (AI) email marketing. The automation capability is dependent on the deep learning and machine learning capabilities of the chosen platform. Businesses are reducing their workload using this capability. There are templates available across AI platforms that are increasing the convenience factor of email marketing. Automation can be specific to the subject lines of the emails. Others can automate the content in the form of templates that businesses are using to personalize emails sent to potential clients.

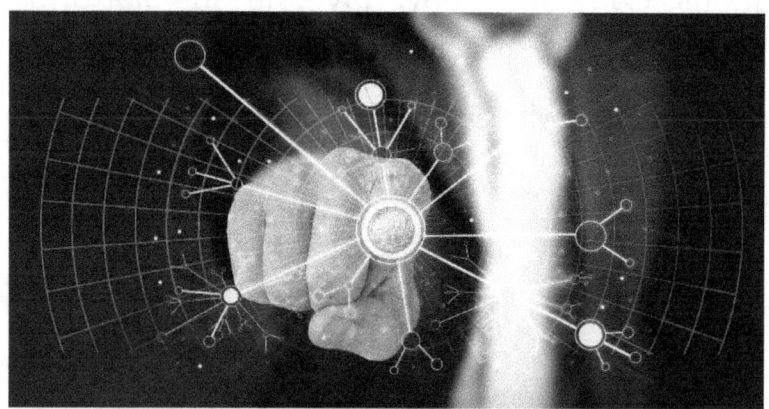

Personalization can be in different forms depending on the goals of the business. Some businesses are opting to personalize the emails by including the name of the intended recipient of the communication. A company can specify the time of receipt of the email. Depending on the platform in use for AI email marketing, a business can choose to align the affiliate links within the email message to the interests of the customer. Some are opting to include promotional messages within their emails. These can act as lead generation opportunities with the goal being to interact in a human-like manner with clients.

The convenience of automation regarding AI email marketing is allowing for the capability of optimization. Data is driving the optimization with the platforms' sourcing for information based on machine learning and deep learning capabilities. Optimization can be of capacity in handling more customers. Segmenting of customers through AI-powered email marketing is contributing to optimization. Businesses can take into account the effect of local timings when sending emails to potential clients. AI is helping in the optimization of personalization as one does not have to manually insert, for example, the name of the recipient per email sent. Optimization of content is now possible as businesses can link email triggers to customer moods.

The convenience of automation regarding AI email marketing has the advantage of frequency setting. Businesses can set intervals at which potential clients are to receive their emails in bulk with little labor expended. Details like how many emails require sending at any one time are now possible. Businesses are setting these details in advance, allowing for aligning to strategic goals for periods at a time. The convenience of automation can take into account the dynamic nature of the marketplace. Some businesses are opting to align their frequency schedules with the actions of their consumers.

The goals of the convenience of automation regarding AI email marketing are varied. These may include conversion for businesses looking to move potential customers to make purchases or take action like subscribing to, for example, newsletters. Some companies are looking to minimize their manual workload. The convenience factor is increasing engagement for businesses looking for an interactive relationship with their clients. With convenience, some organizations want to increase sales. The increase in sales might translate to increased income for the businesses. Some want to serve their customers faster to gain a competitive edge.

Some challenges come with the convenience of automation in AI email marketing. These may include, for example, the ability to align with the context in which customer segments are receiving the email. An email may have been pre-programmed when dynamics in the marketplace were different. Depending on the complexity of tasks required by a business regarding AI email marketing, the costs of the platforms may increase. The charges and impact should be balanced. With more companies providing platforms for AI email marketing, businesses may get confused about which one would best fit their dynamics. Organizations can use marketplace reviews as a filter.

Why Apply Artificial Intelligence to Email Marketing

The reasons why businesses should apply artificial intelligence (AI) to email marketing vary. The context of the industry within which the organization belongs can determine how applicable AI is to email marketing. Some industries are more conservative than others. There are industries where AI in email marketing is a standard expectation by potential clients affecting their perception of businesses. AI in email marketing is allowing companies to track email metrics. The aspects may include time length before an email opening and the percentage of those opened. Some businesses are achieving an increasing return on investment (ROI) with AI email marketing.

Personalization is achievable with AI email marketing with less manual labor. Businesses can choose to personalize the subject line for various reasons, like capturing the attention of potential clients. AI allows companies to automate the personalization of the addressee within the body of the email. The content of the email can be in alignment with the end goal of a company with tweaking done to fit the communication context. Information shared can be in alignment with client needs. The aligning has its basis as the data mining that is possible with AI-based email marketing.

Businesses are achieving a competitive edge in the marketplace using AI email marketing. The promotions within AI emails are allowing companies to predict the needs of potential customers with higher accuracy. These marketing communication channels can adapt to the dynamic nature of customers. Faster responses are now possible with AI email marketing, therefore, improving the customer experience. Conflict resolution is achievable quicker in case of miscommunication. The opportunity to include FAQs (Frequently Asked Questions) within email marketing is portraying businesses as responsive to potential clients. Interactive emails through AI technology can increase brand loyalty. Data generation in AI email marketing is from the capability of machine learning within the platforms.

AI email marketing can take into account the dynamics of differences in global time. During pre-programming, the timing of an AI email message can be in alignment with variances to the appropriate time of delivery. A business would not want an email with a morning greeting arriving in the evening. These tweaks can determine how potential customers perceive the image of a company. The timing of an email can affect the impact of the information shared. Global timing is critical in the context of culture with marketing best if aligning with the expectations of potential customers.

AI email marketing can help businesses capture the attention of potential clients. First, businesses can align with client expectations within the industry regarding how responsive they are to potential customers. Automated emails due to AI give companies a professional outlook. AI email marketing is allowing businesses to have a trackable system of follow up for customers who reach out to them. Organizations can track multiple customer segments consecutively. AI email marketing is giving customers a choice to connect with businesses shortening the sales cycle as, for example, subscribing denotes interest from a potential customer.

With AI email marketing businesses can achieve segmentation of customer groups. Companies can use the segmentation to rethink the promotional messages within their AI email message templates. The segmentation is allowing companies to gain more insight into client groups. Some companies are manufacturing new products and services from data analysis of the customer segments, which in turn is increasing their market share in the marketplace. Segmentation can be in alignment with the culture and time zones. Businesses can add new segments according to the changing interests of their potential customers in real-time, depending on the capabilities of the AI platform in use.

Efficient Send Time Optimization

Businesses using artificial intelligence email marketing should consider the effect of time. Globally, the world has different time zones with some countries experiencing the same within their borders. An email, for example, sent in the afternoon in one state, though sent and received immediately may be at night in another region. The impact of the body of an email message is affected by these variations. An example is sending an email message that a business intends to send best wishes to a client on a particular holiday, yet it arrives a day later. The timing effect, in this case, negates the impact and intent.

Various platforms provide different methods of efficient send-time optimization. Businesses would want to choose platforms that allow for the segmentation of customers by time zones regardless of their locations worldwide. The platforms allowing for optimization up to the second component are preferable. Some time zones differ with days, some hours, and even down to minutes. When thinking of efficient send-time optimization, one should consider the challenges of time zones heading to weekends. AI technology is critical in achieving this capability across platforms in a convenient manner. This optimization allows businesses to achieve their goals.

When establishing efficient send-time optimization, businesses should consider the location of the intended recipient. Issues like culture can determine the appropriate time for sending email to achieve the intended impact. Time zone consideration is a critical factor when looking for efficient send-time optimization for emails to recipients. The focus is on the time the recipient is to receive the communication. Businesses should choose the time that has the best impact concerning the intended message the company is looking to communicate. The content of the email message can be an indicator of the optimum time of receipt.

The benefits of efficient send-time optimization to businesses are many, including receiving customer responses in real-time. Near-instant responding can help companies in adjusting to the dynamic nature of the marketplace. Quick turnaround time is known as a loyalty-building quality among customers. Some businesses are opting to outsource this element of email marketing. The optimization is creating an opportunity for organizations to improve their marketing strategies.

Data from efficient send time optimization responses can give new insights to businesses. Some businesses have created new product and service offerings from the analysis of such data. When factoring in time zones, companies should consider downtimes and use the same to their advantage.

The process of achieving efficient send-time optimization has its challenges. Businesses must wade through the myriad of platforms claiming to offer artificial email marketing tools that can attain the goal. Companies can opt to rely on reviews from the marketplace. The challenge comes in for a new provider with apparently unique possibilities which may determine whether a company moves to the top of the market. The choice may involve a risk that may lead to a loss for a business. When choosing AI tools for achieving efficient send-time optimization, companies should consider whether the cost of implementation is worthwhile.

Businesses have no choice but to adapt to efficient send-time optimization to keep up with dynamic marketplaces. Customers are demanding faster and more accurate customer service experiences. Companies that are not aligning with customer interests are losing market share.

Some businesses that were once market leaders no longer exist due to their slow reaction to market demands. Customers are expecting personalization of experience even to individual levels. Efficient send -time optimization is a tool that businesses can use to differentiate themselves in the marketplace. The capabilities of AI platforms in achieving the same is increasing.

Effective Email Automation Workflow

Having effective email automation comes with benefits that businesses can utilize. It allows for convenient work processes that can give companies a competitive edge within the marketplace. Clients have a positive perception towards organizations with an email automation workflow that is effective. Such an organization has an increased capacity. Customers consider such organizations as responsive. Effective email automation portrays synergy within business processes to the benefit of the customer. Businesses can use the automation workflow to meet customer needs in real-time.

Effective email automation workflow has its base as an AI. The AI activates the automation process. An effective workflow focuses on following up on the details that the business is using as triggers for the automation process. The result is the human workforce can focus on core business processes. There are visual workflows. Companies use the optical element to have an overview of the workflow at a glance. The email automation workflow builders are allowing businesses to see both the email automation and flow in tandem. There are a variety of platforms currently providing these builders. Access for interested companies is possible. The builders make the automation process simple. The advantage is they save time for businesses looking for a convenient way of achieving effective email automation workflows.

It is possible to achieve an effective email automation workflow. Businesses can opt for platforms that are providing visual builders over the traditional manual approach. Integration into the AI system of the triggers for realizing an effective email automation workflow is possible. New triggers can be identified by AI systems increasing workflow effectiveness. Businesses can use the new triggers to communicate better with potential clients. Nurturing of leads is possible. This capability is allowing companies to keep pace with the changes within the marketplace. Visual elements in building effective email automation workflows work well for those who are visual learners.

Coming up with an effective workflow can be taxing. Those who are not optical learners may not appreciate visual workflows. Some platforms providing email automation workflows may not be versatile enough to integrate with other applications in use by businesses. As with all software, AI-based systems are prone to hackers. Some AI software may be susceptible to computer bugs. Some individuals may negate the advantage of automation by over detailing email automation workflows. Some AI systems in use to back the email automation workflows may be costly vis a vis the expected impact.

Businesses should institute email automation workflows. Customers today are more demanding of quality services which may entail responding to potential clients. Companies that choose to ignore the demands of potential clients may lose their share in the marketplace. An organization can no longer afford to react slowly to customer inquiries. The result would be negative news. Customers today can spread a negative experience at a faster rate than in yesteryears. Email automation workflows are allowing businesses to take care of all critical tasks in the process of communicating with potential clients.

There are various platforms providing email automation workflows. Businesses should opt for those that are in alignment with their strategic objectives. The implementation of the workflows using platforms that are in alignment with the strategic goals will have a better impact in the marketplace. Businesses can opt for platforms that have good reviews. Experience can be a critical factor in providing the right support in case the company faces any challenge integrating the email automation workflow. The cost versus ROI (Return on Investment) should be a consideration. The user interface should be easy to use for a smoother integration process.

Chapter 9: Exploit Artificial Intelligence of Big Companies

Exploiting the artificial intelligence of big companies can be the foundation upon which a business builds its presence in the marketplace. Some companies have experience in the AI industry spanning decades and are improving their platforms continuously in a bid to grow their market share. These companies offer opportunities to smaller establishments to ride on their achievements as they move towards their strategic objectives. Small companies can choose a big organization aligning with their interests.

The advantages of exploiting AI of big companies are many, including the opportunity to create a competitive edge in the marketplace. These companies already are attracting brand loyalty and in extension can cause their clients to trust a small business seen to be sharing their AI system. When using the AI of big companies, small businesses may not have to worry about cybersecurity as much. The big companies are continuously looking for ways to protect their systems, therefore, having better protection than small businesses.

Big companies that have AI that small businesses can exploit are many with their platforms having different benefits that may assist small businesses. Some of the companies include Amazon, Apple, and Facebook. Another big company having AI that other organizations can exploit to their benefit is Google which has many AI products currently in the market. Each of the big companies produces AI platforms that are helping them in achieving their strategic objectives.

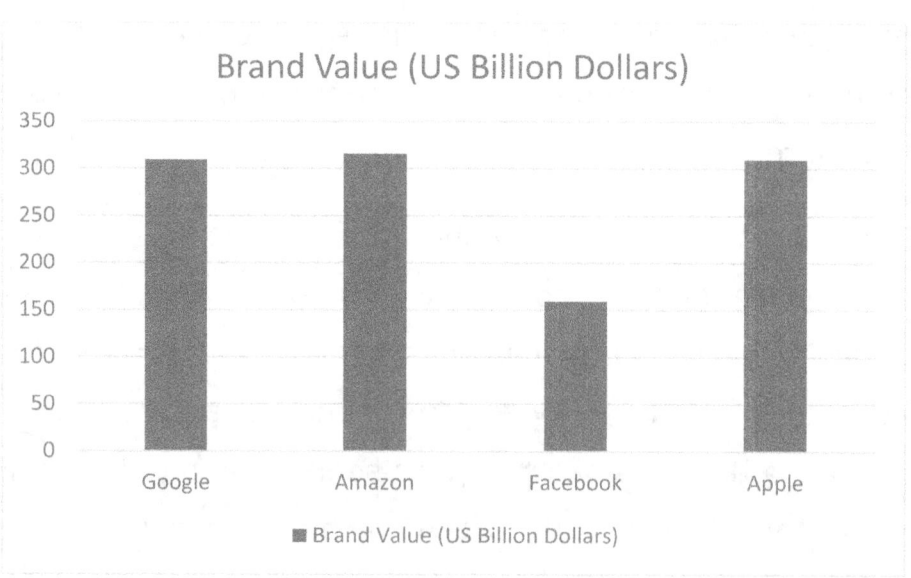

Google's Artificial Intelligence Activities

The artificial intelligence activities of Google are many with the company investing in continuously improving their AI capabilities which are in use across many businesses. These include Google Ads, Google Maps, and Google search engine, which companies are using worldwide. Each of these Google tools is providing varied benefits to those exploiting their capabilities within their contexts. Google's AI platforms are generally accessible online with different tools geared towards specific population segments. The abilities of Google AI-based activities depend on the generation and analysis of data. Machine learning and deep learning are capabilities of these AI platforms with new versions periodically released.

Google Ads is an AI activity for marketing and information purposes with the platform allowing businesses to create personalized ads. The delivery of these ads is dynamic with potential clients seeing them when they align with their changing interests. As potential customers are browsing through websites in affiliation with the Google company, they come across ads relevant to them. Currently, the AI capabilities regarding Google Ads are allowing for a suggestion of how to create ads making the process more convenient. The creative ad capability is allowing Google Ads to act as dynamic templates that businesses can exploit to further their interests.

Google Teachable machine is an AI that is for the student population, making it a platform that can expand the opportunity for businesses in the education industry. Companies looking to explore opportunities in understanding the concept of machine learning can find this AI useful. The teaching process has not integrated coding, which makes it easier to use for those inexperienced in computer languages. One uses a camera to aid the machine in learning with the lessons' facilitation being online-based. The platform is combining the advantages of AI and machine learning.

The Google Search engine may be the most recognizable AI from the company with businesses using the tool to source for information. The search engine uses algorithms that trigger answers to searches that come up when users type in keywords when looking for information. This AI platform can specify search results based on the location of the one typing the keyword giving greater information accuracy to businesses. Language-based results are possible, which creates a better customer experience for potential clients. Google search engine is now allowing users to translate pages that can create a connection with potential clients as they feel a part of the company.

Google Maps is an AI in use by businesses to augment their operations, for example, by the taxi-hailing apps available in the market. The drivers can use the AI to determine their location and the directions to where they intend to go, including tips like the shortest route. Google Maps are helping brick and mortar stores showcase their position, which may lead to an increase in walk-in clients. Companies can use Google Maps to increase customer interactivity by establishing tracking maps allowing customers to see the workflow of their transaction.

Regarding the AI technology by Google in use in search engines, businesses can use voice options to triggers searches to topics relevant to the organization. The Google AI of choice for companies can focus on needs the company is looking to fulfill both within and without the business entity. Businesses can employ varied Google AI platforms, consecutively. The Google AI possibilities are allowing for integrations with existing applications in use within organizations. There are specific Google AI applications geared towards assisting businesses to increase efficiency in various processes. The processes the AI tools may cover include marketing, sales, and operations.

Facebook's Artificial Intelligence Activities

The AI activities of Facebook can be thought to revolve around the underlying social nature of Facebook, taking into account the social changes. Businesses can use Facebook AI to track and report the dynamic changes within the society that may affect their operations. The follow up is via deep learning and machine learning with the details being of the highest quality. Facebook is generating purely social AIs that businesses can take advantage of the opportunities it presents. Facebook's AI activities involve more than one billion people giving their data a high level of data, and therefore higher levels of accuracy.

Facebook AI activities include analytics which businesses can use in understanding the interests of their clients and aligning their product and service offerings to their needs. The analytics Facebook shares are specific and dynamic as the data in use is continuously updated as people are interacting within their platform. Businesses can determine the parameters that they want to see, for example, filtered through location, age, and distance. Organizations are using the data from Facebook analytics to create new product and service offerings. The use can propel a business to a position of influence within the marketplace.

Chatbots are now available within Facebook for both personal and business pages with the latter helping organizations take advantage of the benefits of the AI tool. These are for messaging clients in a timely fashion using the automatic nature of AI-powered chatbots. The chatbots are in use by businesses for different purposes, for example, gathering of customer information. Chatbots by Facebook are in use by companies to upsell and market promotions to their potential client bases using the platform. Businesses are including links within the chatbots. These are leading customers to their websites, hoping to turn them from interested customers to actual clients.

Sponsored ads by Facebook is another form of AI businesses can take advantage of within the platform. These require payment by the organization after a set period. Companies are using ads to reach their intended target clients. The timing of the sponsored ads is possible, allowing a business to personalize their customer experience, which can be a source of the competitive edge in the marketplace. The types of ads companies can create through Facebook are different, including videos, and photos which can help a business reach visual customers. Facebook shares data on how the promotions are performing and allow changing of the content of the ads as they are running.

Facebook AI activities include running of survey questions within a user's newsfeed. The surveys are generally short containing one or two inquiries that require a click response which encourages a higher rate of customer action. Businesses can use this feature to get feedback from their potential customers regarding product and service offerings. The survey feature offers convenience as companies can create their own from their Facebook business pages. Businesses can create surveys in different forms, including poll questions. They can be versatile in terms of visual presentation, for example, including videos and pictures.

Facebook AI allows for third party integration of applications making it a versatile platform that businesses can align with their interests. These can be affiliated directly to Facebook like Instagram or not, for example, Survey Monkey. For Facebook affiliated programs, ad sharing is possible, which makes its use convenient. The integration can allow for smoother transition with applications already in use by businesses in various stages of their internal processes. It can save time for companies that have a presence across different social media platforms with their activities appearing on Facebook and other applications simultaneously.

Amazon's Artificial Intelligence Activities

The AI activities from Amazon are many with their focus being on continuous improvement of their systems that are giving them a competitive edge within the marketplace. The actions have grown it into a behemoth of online trading with a wide variety of products that potential customers have access to through their accounts. Amazon AI activities can benefit businesses of different sizes as their platform is versatile, allowing both digital and physical products. The actions have their basis in deep learning and machine learning which helps Amazon consistently improve on their offerings to potential clients. The developments are helping in breaking down barriers that were due to geographical limitations.

For physical products, Amazon has an AI-backed tracking system that allows customers to have real-time information on the location of their packages. The transparency has built their reputation as trustworthy, which is giving them a presence worldwide in terms of trade. Businesses can tap into this brand trust to move their items through the platform of Amazon around the world. Their experience with shipping is unmatched, which can save smaller businesses from dealing with the challenges that come with logistics. Companies can outsource tasks that Amazon can handle to focus on their core competencies, leveraging on their AI activities in the marketplace while achieving scale operations quicker.

The AI activities by Amazon span the world of payment processing with transaction confirmation occurring via automated email processing. The perception in the market of their ability to handle payments without fraud is positive enabling businesses to reach clients that trust its business model. Plugging into their robust system may mean growth for organizations as it exposes companies to new customer segments at a low cost. The AI activities at Amazon are allowing for cost calculation in a customer's local currency, making the transactions relatable. Businesses can, therefore, forecast their earnings in a currency that fits their local contexts.

Using AI, Amazon is allowing for the creation of personalized stores both for customers and for businesses which creates a better customer experience. One can control their purchasing or store collection conveniently with buying done from the comfort of one's home. Companies can lead potential customers to their storefronts within the Amazon platform at no extra costs to their business. Organizations can change the look and feel of their digital storefronts conveniently with the touch of a button, which they can use to attain a competitive edge within the marketplace.

Amazon is using robots as part of their AI activity strategy. With the machines assisting the human workforce in varied business processes, they can achieve a quick turnaround time. Companies can leverage this advantage to reach customers faster while saving on costs. One of the areas where Amazon is integrating robots is within its warehouses to reduce errors and increase the speed of service. Businesses can use this strength to improve their customer experience, which may convert to increased market share. This investment is allowing Amazon to compete effectively with brick and mortar stores.

Amazon's innovation within the field of AI is allowing it to compete within the digital marketplace with the company creating its digital bookshop known as Amazon Kindle. Here, clients can access millions of books worldwide at the touch of a button which improves customer experience. Businesses can, therefore, go around the limitations of geographical barriers, which may increase their market reach. Uploading digital products is also convenient with individuals globally able to benefit from the technology. Amazon's AI activities, in this way, levels the field for businesses of smaller size to compete with the larger establishments.

Apple's Artificial Intelligence Activities

Apple's AI activities in the technology space are revolutionary with the company having a segment of customers who portray brand loyalty through their purchases. They run on the foundation of exclusivity with offerings of products and services available to those who select to purchase their devices. Data and machine learnings are part and parcel of the innovations within the company's portfolio of products. Their segmentation of customers can be a market source for businesses looking to serve their exclusive club. Such companies will not have to bear the cost of segmentation as Apple takes on the same.

Apple, as a company, achieves exclusivity through the process of encryption. It does not allow non-affiliated systems to work with their products. The result is improved sales as customers have to keep updating their devices to remain in the exclusive club and the improved quality of services for their customers. The latter is through the passing of quality standards set by the company to be on their platform. Their customers also trust them giving them the advantage of brand loyalty that businesses can tap into to sell their products. Given the characteristics of their customer segments, organizations can tailor their product promotions to align with the interests of their clients.

Apple AI activities inform their sales strategy with precise segmentation, for example, aligning with expected income levels of potential customers. Apple devices are undergoing continuous improvement, establishing a perpetual sales cycle. Customers, will over time, update their devices through purchasing as changes reach a point of not being in tune with a tool. Businesses can align themselves with their sales strategy to come up with new product offerings that can be of interest to the customer segment within Apple's platform. The companies can improve their sales revenue while riding on the business model of the Apple company.

Apple AI activities come with challenges arising from their business model. Some are the attraction of those looking to hack into their system of exclusivity. There are scenarios where there are calls for the exclusivity to align with the political interests of societies. Some customers may not be able to keep up with the device changes in terms of purchasing, which may lead to a loss of customers. Due to geographical limitations, some customers may not access the latest innovations presented by the company to their client base. There has been a criticism of the company's ability to maintain a lead in AI innovation, given its model of exclusivity, which is an expectation of customers.

Apple, as a firm, operates the Apple store that holds different products, which may be of interest to their customer base. The products in the store must pass their standards to be part of the portfolio. Businesses can work to attain their requirements to sell their products and services to the customers, which may expand a company's market share. The exclusivity may work to the advantage of a company as customers may perceive a brand to be exclusive when associated with the Apple company.

The Apple company creates devices that are compatible with one another, and that cannot work with other third-party applications. Companies, depending on their strategic goals, can aim to have their devices work with the exclusivity of the Apple brand as a foray to new markets. Companies can focus on adjustability beyond the Apple Company. The effect could be achieving an extensive array of clients. Customers are now demanding products that are not restrictive which may be an opportunity for businesses. Versatility to a broader arena can mean working with more third-party applications as they would assume the products of a company are of the level of the Apple Company regarding the quality levels.

Conclusion

In the book, *Artificial Intelligence for Business Applications,* the reader will find information describing what artificial intelligence entails. The promises and challenges of the software are covered. There are discussions on how businesses can take advantage of the hopes of AI.

They will learn how AI is changing business processes. The processes discussed include new customer acquisition. One will find details on customer services that are coming up as businesses are implementing AI. The book discusses the topic of virtual assistants in business settings. The author describes how companies are using AI in the context of focus and strategy where the reader will find details on how to plan one's AI strategy.

Business owners can learn how to predict the behaviors of their target customers by reading the book. The book contains discussions on chatbots and responders. One will find extensive descriptions of what they are in the context of AI.

The author shares information on marketing platforms that have AI capabilities. Businesses can find information on how to use AI to improve their email marketing processes with the benefits of the same discussed in detail. The book details why companies should take the idea of applying AI in email marketing seriously.

The book contains ideas on how a business can exploit the AI of big companies. One will find specifics on big companies providing AI and how they can take advantage of the same. The book details the type of AI the companies are currently running on their platforms.